CALEB SCHARF

Gravity's Engines

The Other Side of Black Holes

PENGUIN BOOKS

PENGUIN BOOKS

Published by the Penguin Group
Penguin Books Ltd, 80 Strand, London WC2R 0RL, England
Penguin Group (USA) Inc., 375 Hudson Street, New York, New York 10014, USA
Penguin Group (Canada), 90 Eglinton Avenue East, Suite 700, Toronto, Ontario, Canada M4P 2Y3
(a division of Pearson Penguin Canada Inc.)
Penguin Ireland, 25 St Stephen's Green, Dublin 2, Ireland
(a division of Penguin Books Ltd)
Penguin Group (Australia), 707 Collins Street, Melbourne, Victoria 3008, Australia
(a division of Pearson Australia Group Pty Ltd)
Penguin Books India Pvt Ltd, 11 Community Centre,
Panchsheel Park, New Delhi – 110 017, India
Penguin Group (NZ), 67 Apollo Drive, Rosedale, Auckland 0632, New Zealand
(a division of Pearson New Zealand Ltd)
Penguin Books (South Africa) (Pty) Ltd, Block D, Rosebank Office Park,
181 Jan Smuts Avenue, Parktown North, Gauteng 2193, South Africa
Penguin Books Ltd, Registered Offices: 80 Strand, London WC2R 0RL, England

www.penguin.com

First published in the United States of America by Scientific American/Farrar, Straus and Giroux 2012
First published in Great Britain by Allen Lane 2012
Published in Penguin Books 2013
001

Printed in Great Britain by Clays Ltd, St Ives plc

978-0-241-95735-6

www.greenpenguin.co.uk

PENGUIN BOOKS

Gravity's Engines

Caleb Scharf is the director of Columbia University's Astrobiology Center. He has written for *New Scientist*, *Science* and *Nature*, and his *Life, Unbounded* blog was named one of the 'hottest science blogs' by the *Guardian*. He was born in England, and now lives in New York City.

CONTENTS

PREFACE

This is a book all about remarkable science that stretches from highly theoretical descriptions of natural phenomena, arising from the deepest recesses of human thought and intuition, to the most visceral and visible pictures of the real universe. It's a story about physicists and astronomers hunting for black holes, and about our quest to understand cosmological truths, galaxies, stars, exoplanets, and even life on other worlds. Black holes have generated a particular fascination ever since they emerged into mainstream popular culture in the 1960s and '70s. Weird, destructive, time-warping, overwhelmingly alien, they've provided endless fodder for both science writing and science fiction. As astronomers have labored to process torrents of new data and to build an ever better picture of the universe and its contents, they have found not only that black holes are a significant and even critical part of that atlas, but that many are fearsomely noisy and rambunctious. These are crazy, exciting, and challenging discoveries—blockbuster stuff.

The stories in this book highlight just how important I think these objects really are. Black holes are gravity's engines—the most efficient energy generators in the cosmos. And because of that, they have played a key role in sculpting the universe we see today. This is, to my mind, one of the most extraordinary and bizarre characteristics of nature that we've yet stumbled across: some of the most destructive and naturally inaccessible objects in the universe are also the most important. It's well worth pausing to consider this, and I, for one, think the journey is tremendously enjoyable.

Of course, the story I'm telling here relies on the clever and exhaustive work of many, many outstanding scientists. Their collective contributions have inspired and influenced my own thinking. What I'd like you to take away from *Gravity's Engines* is both a sense of the cosmic grandeur we have discovered and a feel for the great scope and ingenuity of human ideas at play. If you still find yourself hungry for more, the notes will give you a taste of the vast ocean of literature from which I've fished the choicest morsels. If nothing else, skim through the end matter to get a sense of the incredible wealth of human thought involved.

I find writing about science to be a fascinating experience. After spending a fair chunk of my life *doing* science, turning around to construct a story *about* science is both illuminating and humbling. I am indebted to many sources for facts and inspiration. Books by Kip Thorne, Mitch Begelman, and Martin Rees deserve special mention. I've found these and many other works to be essential along the way, and they appear in the notes at the end of my book.

Many others deserve thanks for a multitude of reasons. For writing: This book would be vapor in the aether but for the efforts of my wonderful and insightful agent, Deirdre Mullane of Mullane Literary Associates, and the hard work and extraordinary skills of Amanda Moon at Scientific American / Farrar, Straus and Giroux, who has guided me with graceful patience throughout the process.

For science: This whole enterprise really began twenty years ago with two important mentors. Ofer Lahav and Donald Lynden-Bell generously shared their wisdom and helped foist me on the world of professional astronomy. Along the long subsequent path I'd like to thank Keith Jahoda, Richard Mushotzky, Laurence Jones,

Eric Perlman, Harald Ebeling, Donald Horner, Megan Donahue, Mark Voit, Andy Fabian, Keith Gendreau, Eric Gotthelf, Colin Norman, Wil van Breugel, Ian Smail, David Helfand, Mark Bautz, Frits Paerels, Steve Kahn, Fernando Camilo, Francisco Feliciano, Nelson Rivera, Arlin Crotts, Zoltán Haiman, Joanne Baker, Michael Storrie-Lombardi, David Spiegel, Kristen Menou, Ben Oppenheimer, Adam Black, Mbambu Miller, Greg Barrett, Jane Rosenman, and many others who through no fault of their own have inspired and encouraged me.

For all the other things: To my personal cheerleading squad and long-suffering family—matriarch Marina Scharf, wife Bonnie Scarborough, daughters Laila and Amelia—I owe you pretty much everything.

Finally, a little thought before you begin reading the book. As a species we are born out of 4 billion years of fierce molecular evolution that leaves us eager to work and work and work. We do it to survive, and for far too many of us that survival is still not guaranteed. For others it is more a means to an end, a way to provision ourselves with the things that bring comfort, joy, and even some peace. Nonetheless, we should all consider taking just a moment now and then to stop and gaze skyward. As tiny as we are, our lives are tied intricately into an amazing and grand cosmos. This is our heritage. We should be proud and satisfied with our place in it, and never put aside our curiosity about it.

GRAVITY'S ENGINES

1

DARK STAR

A computer sits among the coffee-stained papers scattered across my desk. Its screen has been blank all morning. Suddenly it lights up and displays a pixelated image. A message is coming in from space.

A few days earlier, high above Earth's surface, a great orbiting observatory has stared for forty hours over the bows of the Milky Way galaxy. With chilled eyes it has patiently tracked a tiny patch of the cosmos, a speck of sky close to the constellation Auriga—the Charioteer. In this direction is a glorious view for a spotter peering into the abyss in the hope of finding treasure.

This remarkable instrument is called Chandra. Decades of work went into its construction, with hundreds of people toiling in multiple countries. The blood, sweat, love, and tears of a highly technological civilization produced the smooth surfaces and exquisitely precise devices inside it. Careers started and ended while it grew from a dream into a reality. Finally it was lofted into space and released with tender delicacy from the belly of NASA's space shuttle *Columbia*, becoming a tangible example of humanity's endless curiosity.

Now it has captured a whiff of something from the deep. Photons, particles of light, have found their way down through its mirrors and filters, forming an image on the silicon sensor of a digital camera. That image, encoded as a stream of data, has passed to Earth, first beamed as microwaves to a ground station and then relayed around the globe. Processed and sent on across a continent, another journey through hundreds of miles of wires and fiber optics, it finally re-forms as a monochrome picture on a screen in my small and untidy office ten floors above the streets of twenty-first-century Manhattan.

On any given day, we don't expect to see much that is particularly remarkable in the vast flood of incoming data that is a part of modern science. Patience is a hard-won lesson. Yet there, amid the rough noise of the image, is a structure. It's small and faint, but unmistakable. I can see a pinpoint of light surrounded by something else—a fuzzy streak jutting out to the left and right. It looks like a small dragonfly pinned to a piece of cardboard. Something is very curious about this image. It has the flavor of a new species.

Traffic out on the street echoes noisily up the canyon of buildings, but for an instant it rings hollow. My mind is not in this world anymore, but away in a very, very distant corner of the universe.

Twelve billion years ago, the photons that made this image began their journey. They are X-rays, invisible to human eyes, but able to penetrate through our soft bodies. For 12 billion years they have passed unimpeded through the cosmos. But as they have traveled, the universe has changed; space itself has expanded, stretching the photon waveforms and cooling them to a lower energy.

When they set out there is no star called the Sun, no planet called Earth. It isn't until they are two-thirds of the way through their journey that part of a collapsing nebula, a cloud of interstellar gas and dust in a still impossibly remote galaxy, produces a new star and a set of new planets that will eventually become our home.

When Earth forms, these photons are already ancient, 7-billion-year-old particles that have traversed vast stretches of the cosmos.

Time passes. Somewhere on Earth a complex set of molecular structures begins to self-replicate: life begins. Two billion years later, the photons start to enter the very outer regions of what we might call the known universe. Here are the great superclusters and web-like structures of galaxies that we have mapped. Spanning tens of millions to hundreds of millions of light-years, these forms are the skeletons upon which galaxies and stars are coalescing, molded by gravity—millions of galaxies, and quintillions of stars, strung through the cosmos. On Earth, microbial evolution has just given rise to the first cells of a new type of life—the Eukarya, our direct ancestors. These busy microscopic creatures swim off in search of food.

A billion more years go by. The photons enter truly known space, a realm where our instruments have mapped great walls of galaxies and huge empty voids. Here are structures with familiar names and calling cards, like Abell 2218 and Zwicky 3146, huge gravitational swarms of galaxies known as clusters. On Earth the very first true multicellular life emerges, and the air is filling with oxygen. The chemistry of this element is ferocious. New types of metabolism are evolving in response—a revolution is under way. Just 500 million years later, the dry surfaces of Earth are covered by something exotic: plants that use the molecular tools of photosynthesis. A strange, greenish tint appears across the supercontinent Gondwana, the largest body of land on the planet.

The photons continue their patient journey, passing through regions that will be increasingly familiar to as yet unevolved astronomers. Nearby are the great galaxy clusters we will name for the constellations in which we see them: Coma, Centaurus, and Hydra. Onward the photons fly, and from the point of view of an observer standing to the side as they race past, our galaxy is now one of thousands of patches of light in the sky ahead.

It takes them another 490 million years to reach our Local Group, a ragtag band of galaxies. Some are large like Andromeda and the Milky Way, and some are small, like the dwarf galaxies Cetus, Pegasus, Fornax, and Phoenix. It is not a particularly remarkable place, perhaps a total of a few trillion or so stars altogether.

On Earth many great periods of life have come and gone. The dinosaurs have not been seen for almost 60 million years. The continents and oceans have changed dramatically, and the contours of our modern world are clearly visible. Birds and mammals are swarming across the globe. The Black, Caspian, and Aral Seas are beginning their separation from the ancient Tethys Ocean and what will become the Mediterranean Sea.

In the next few million years, the photons descend into the gravity well of our galactic neighborhood. The Milky Way is now a distinct glowing smear reaching across the sky as it gets ever closer. On the third planet from a modest G-dwarf star orbiting in one of the outer arms of this spiral galaxy, a new type of animal begins to walk upright on two legs. As it leaves its footprints in muddy volcanic ash in what is now the Olduvai Gorge in eastern Africa, the photons speed ever closer. Now in almost their 12 billionth year, they have never slowed down. As particles of light they are threaded into space and time, moving at the same constant speed as they did at their origin.

It takes them another 2 million years to reach the outermost wisps of our great Catherine wheel of a galaxy. A major glacial age is taking place on Earth. Huge ice fields grow outward from the planetary poles, engulfing the northern hemisphere. This profound change in environment impacts the behavior and fortunes of the hominids' descendants—humans. Groups of people migrate and explore. Areas that were once shallow seas are now traversable on foot. Another twelve thousand years pass and the photons fly in across the spiral arm of stars, gas, and dust in the Milky Way galaxy that is called Perseus. By now the ice has retreated, and new pockets of humanity are scattered far and wide. Great cultures have risen and vanished, and others are beginning to flourish across the planet from the Middle East to Asia, from Africa to North and South America, and in Oceania.

The photons enter the Orion spur of our galaxy. To one side they pass the Orion nebula itself, a vast and beautiful cloud of gas and dust, the birthplace of new stars and the graveyard of old ones. One thousand years remain for their great migration. On Earth, Chinese and Middle Eastern astronomers observe a new bright object in the heavens. Unknown to them, they have witnessed a supernova, the explosive death of a star. A decade later, in the year we now label 1066, a duke from Normandy ingloriously named William the Bastard leads his army in the conquest of an island kingdom where he claims the throne. Preceding his arrival, and believed to presage it, a glowing comet, later to be known as Halley's, passes through the skies and is depicted in the epic Bayeux Tapestry recording these great events. This is the nineteenth time that it has been documented by human observers, each sighting some seventy-five years apart.

Kings and queens, emperors and empresses rise and fall. Wars flare up and eventually end. Humans migrate and explore the planet. Diseases, volcanoes, earthquakes, and floods ebb and flow as time goes by. Six hundred years pass in the blink of a cosmic eye. The photons enter a sphere centered on Earth that encompasses the Pleiades star cluster, the Seven Sisters. The Sun is a nondescript point of light in the distance. Galileo uses a telescope to study the moons of Jupiter, realizing that they orbit that body, and therefore Earth is not the center of all celestial paths. Half a century passes and Newton formulates physical laws that describe the properties of motion and of gravity.

The photons continue on through the great emptiness of interstellar space—far more vast compared to the sizes of the stars than intergalactic space is compared to the sizes of galaxies. Hundreds of years pass. World Wars I and II ravage the northern hemisphere of the planet. The photons begin to pass through the collection of stars that form the constellation of Auriga, as seen from the vantage point of Earth. The Vietnam War is flaring up and the Beatles are playing on every radio. *Apollo 8* orbits the moon and, for the first time, human eyes see Earth rising above a new horizon.

Decades later and the photons race in through the outskirts of the solar system. Zipping through the magnetic skin of the heliopause—where the Sun's influence gives way to that of interstellar space—they have just hours to go. Finally, as if playing their part in some great cosmic tragedy, they are captured within a cylinder that is only four feet across, a mere 0.0000000000000000001 percent of the diameter of the Milky Way galaxy within which it is embedded. Instead of sailing on to infinity, the photons are caught in the high orbit of planet Earth, inside the great Chandra Observatory, where they are coaxed deep into a series of nested tubes of iridium-coated glass. In the next few nanoseconds these ancient photons of X-ray light finally encounter something in their path in their long journey through the cosmos: a piece of meticulously prepared silicon, itself composed of atoms that were forged inside another star, dead for billions of years. The silicon absorbs their energy and, where each photon lands, releases electrons into the microscopic pixels of a camera. Within a few more seconds a voltage automatically switches on, sweeping these electrons off to the side toward a line of electrodes—like a croupier gathering up the chips on a roulette table. Here, after a journey of 12 billion years, the photons are registered as electrical charges and converted into something new. They have become information.

On the screen in my office in New York this data creates an image. It is a unique and revealing fingerprint of intensity and energy. Here are the signs of a young and extraordinarily massive black hole, ferociously tearing matter apart in the skies of a distant and now ancient galaxy. Its hunger is extreme and violent. But something new and unexpected is revealed as well. A grasping presence extends further, pushing, molding, and altering the surrounding universe. Dragonfly wings of light jut out around the brightest part of the image where the black hole lurks. Their true scale is staggering: they are hundreds of thousands of light-years across. Their true brightness is immense, representing an energy output a trillion times greater

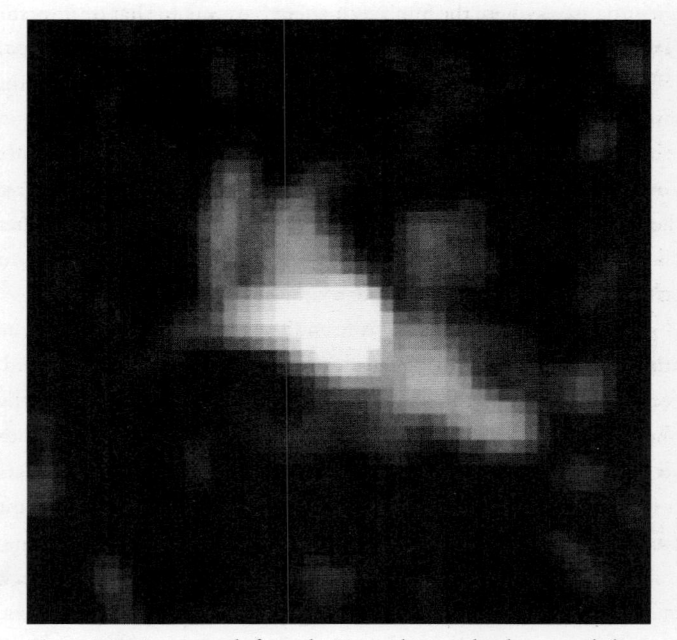

Figure 1. An image made from the X-ray photons that have traveled toward us for 12 billion years. While this picture may look pixelated, this is exactly how it appears at the very limits of instrumental resolution. A bright spot is surrounded by a strange form: dragonfly wings that themselves span hundreds of thousands of light-years. It is a glimpse of a mysterious colossus in the depths of the cosmos.

than that of our Sun. They are flooding that ancient galaxy with radiation, somehow powered by the monster in the middle.

This book is, in part, about the story of this distant place. In the past few decades a remarkable and strange picture has emerged. It extends far beyond the esoteric and fantastic studies of the extremes of space and time that have been a hallmark of black hole science. Astronomy in the late twentieth and early twenty-first centuries has revealed that black holes are both varied and common. While we think that most originate as comparatively small objects, with the mass of just a few Suns, some have managed to grow far larger. The biggest are now known to be *tens of billions* of times the mass of our Sun. They stagger the imagination and challenge our core ideas about how all objects and structures that we see in the universe have come to be. At the same time, they do not hide away as inert bodies, invisible and aloof. We have come to realize that the science of black holes is very real and very immediate. Their presence is acutely felt across the cosmos. Black holes play a critical role in making the universe appear the way it does.

Because of this, they also profoundly influence the environments and circumstances in which planets and planetary systems are formed, and the elemental and chemical mixes that go into them. Life, the phenomenon of which we are a part, is fundamentally connected to all these chains of events. Saying that black holes have implications for life in the universe may sound outrageous and far-fetched, but it appears to be the simple truth, and we are going to follow that tale.

To begin to explain the epic cataclysm appearing on the screen in my office, I have to turn back the clock a couple of hundred years, back to a time when this small armada of photons was still streaking past the outer reaches of the Orion spur of the Milky Way galaxy. Here on Earth it was a different era, one of great change and new ideas—especially in one small corner of the planet.

With its tall and rather austere stone tower, St. Michael's parish church in the village of Thornhill in West Yorkshire, England, seems an unlikely place to nurture the secrets of the universe. Perhaps, though, there is something in the surrounding rugged green terrain or the harsh winter skies that might compel you to wrap up tightly with great cosmic thoughts. Indeed, in 1767 a remarkable thing happened in this small community. Into its midst came an extraordinary thinker, a polymath whose mind roved across the vastness of space. He also happened to be the new rector at Thornhill.

At forty-three years old, John Michell was already a highly regarded figure in British academic circles. He had spent most of his life immersed in intellectual pursuits and had risen to the title of Woodwardian Professor of Geology at the University of Cambridge. His interests were diverse, from the physics of gravitation and magnetism to the geological nature of the Earth. Despite his scientific reputation, however, little personal detail is known about Michell. Some records depict him as short and round, an eminently forgettable physical specimen. Others describe a lively and busy mind, someone who had once met with Benjamin Franklin, was fluent in ancient Greek and Hebrew, was a keen violinist, and kept a household alive with debate and inquiry.

What is clear is that a few years earlier, in 1760, while he was a Fellow of Queens' College in Cambridge, Michell produced a study of earthquakes that established him as one of the forefathers of the modern science of seismology. A decade before that he had written a treatise on the nature and manufacture of magnets. He had also written works on navigation and astronomy, from the study of comets to stars. While he may have been short in physical stature, his sharp vision could pierce through the void.

We can only presume that in the relative calm of life at St. Michael's, Michell was able to find a secure income and home for his family. Perhaps it also gave him time to think away from his otherwise busy rounds of scientific debate in nearby Leeds, and from the great changes being wrought on the world around him. The Industrial Revolution had begun in Europe, Catherine the Great ruled Russia, and the American Revolution was gathering momentum to the west. Less than a hundred years earlier Isaac Newton had published his monumental works on the nature of forces and gravity. Science was becoming its modern self, equipped with increasingly sophisticated technological and mathematical might and emboldened by the times.

There was one problem in particular that caught John Michell's attention when he studied astronomy. It was a fundamental and practical one. While it was well understood that the stars in the night sky were cousins to our own Sun, there remained a deceptively simple question that scientists at the time were unable to answer. In our own solar system, it was clear from geometrical arguments that the Sun was vastly larger than any of the planets. This being the case, it was relatively straightforward to use the estimated distances of the planets from the Sun and the time it took them to complete an orbit—their periods—to estimate the solar mass. Newton had shown how. Newton's universal law of gravitation outlined a simple formula that related the masses of two bodies to the distance between them and the length of the orbital period of one around the other. If the mass of the planets was assumed to be negligibly small compared to the Sun, then the timing of their orbits simply revealed the Sun's true mass.

But the question that vexed Michell was not how to measure the mass of the Sun, but how to measure the mass of distant stars. No planets could yet be seen around them to serve as evidence of their gravity. The physical nature of stars themselves was still unclear. Astronomers understood that they were hot fiery objects, an inference from the way we experience the Sun here on Earth, but their true distances would not begin to be known for another seventy years. Nonetheless it was increasingly clear that the Persian and

Chinese astronomers of the Middle Ages had been on the right track in believing the stars to be out in the distant universe and that they obeyed the same physical laws as those of our solar system. To know their actual sizes would help tremendously in divining their detailed nature.

Michell was an incredibly flexible thinker. In the late 1700s the term "statistics" had barely been introduced to science; the basics of probability theory had been formulated about a century earlier. The idea of applying these tools to real scientific questions was in its infancy. Yet, as he pored over astronomical charts and tables, Michell used statistical reasoning to show that the patterns of stars indicated many were not isolated in space. He proposed that some stars must occur in physically related pairs, or binaries. This observation wasn't verified until 1803, when astronomer William Herschel studied the movement of stars. If one could observe the actual orbits of binary stars, then, using Newton's formula, one could estimate their total mass. But in Michell's time such observation was not quite within the grasp of astronomers, and so he had to keep looking for another approach for measuring the mass of a single distant star.

He came up with a tremendously clever solution. A hundred years earlier Newton had proposed that light was made of "corpuscles"—tiny little particles that traveled in straight lines. Michell reasoned that if light was made of these corpuscles then they would be subject to natural forces, just like everything else. Light escaping the surface of a distant star should therefore be slowed down by gravity. In the late eighteenth century the speed of light was already known to be exceedingly fast—about 186,000 miles a second. Even the great bulk of a star like the Sun, Michell knew, would only slow the light down by a small amount. But if that change could somehow be measured, then the mass of the star could be deduced.

On November 27, 1783, Michell brought his ideas together in a presentation to the Royal Society in London. The title of his paper was a fabulous example of circumlocution and hedging: "On the Means of Discovering the Distance, Magnitude, &c. of the Fixed Stars, in Consequence of the Diminution of the Velocity of Their Light, in Case Such a Diminution Should be Found to Take Place in any of Them, and Such Other Data Should be Procured from Observations, as Would be Farther Necessary for That Purpose."

As he presented his work to the Society, Michell made his argument for deducing the mass of a star. His opening logic was simple: "Let us now suppose the particles of light to be attracted in the same manner as all other bodies with which we are acquainted . . . gravitation being, as far as we know, or have any reason to believe, a universal law of nature." The idea appealed to the audience, who were well versed in Newtonian physics, and by all accounts set them aflutter. Light being slowed by gravity was a delightful notion.

Michell's concept was audacious. The recognition that a star or other cosmic object leaves its dirty fingerprints all over the light that we eventually detect coming from it actually represented a huge leap for modern astronomy. The ability to deduce the nature of objects in the cosmos by the analysis of their light is today central to our exploration of the universe. But Michell had even more to say.

An imaginative problem solver, the rector of Thornhill was clearly feeling inspired. His next big leap was to recognize that an object might be massive enough to pull a corpuscle of light completely to a halt as it tried to fly away. With a bit of mathematical juggling, Michell computed how massive an object would have to be to halt light. He did it by turning the question around. If an object fell toward a star from an infinite distance away and reached the speed of light at the point of impact, then the star had enough gravitational might to prevent light from escaping in the reverse direction. If such a star were of the same density as the Sun, he found it would need to be five hundred times bigger in diameter. His neat summary of the situation for the Royal Society audience was clear: ". . . all light emitted from such a body would be made to return towards it, by its own proper gravity."

From his calculations, Michell realized that this meant there could be objects out in the universe that trapped all light coming from their surfaces and were for all intents and purposes invisible. The only way to spot them would be by detecting their gravitational influence on other objects. Such massive objects in Newtonian physics have since become known as Michell's "dark stars."

A decade after Michell formulated these ideas in the sleepy English countryside around West Yorkshire, the extraordinary French mathematician and astronomer Pierre-Simon Laplace was independently reaching a similar conclusion. Born in Normandy, Laplace was a scientific prodigy, and his mathematical prowess quickly elevated him to the higher echelons of French academia. While still in his twenties, he had single-handedly developed mathematical theories describing the stability of planetary orbits and had helped develop modern calculus. He would go on to help pioneer theories of probability and mathematical physics. What were otherwise Michell's dark stars, Laplace termed "black stars," writing in 1796 that "it is therefore possible that the greatest luminous bodies in the universe are on this account invisible."

Although other scientists were intrigued by these ideas, there is no record that Michell and Laplace ever communicated with each other, and the concept would not be fully understood for more than another century. Newton's corpuscular theory of light fell out of favor, as it failed to explain subsequent optical experiments. Laplace even quietly removed his description of black stars from later copies of his epic work *Exposition du système du monde* (*The System of the World*). Today we know that the fundamental assumption behind Michell and Laplace's theories—that light could be slowed by gravity—is in fact wrong. The truth is far more surprising.

Nonetheless, the idea represented a turning point in thinking about massive objects in the cosmos. It was a revolutionary concept that there could be huge objects in space that are entirely hidden from sight. It was even more extraordinary to suggest that the objects that were the most massive and luminous—throwing off the greatest number of photons, or corpuscles, at any given time—might also be the darkest from our perspective. Exactly how revolutionary these ideas were would not be fully appreciated until much later.

›››

Two pivotal events would eventually bring Michell's dark stars back into view. The first of these was to take place in a chilly basement in Cleveland, Ohio, in 1887.

By the late 1800s, remarkable advances had been made in our understanding of the properties of light and electromagnetism. Decades of experimentation had demonstrated that the flow of electrical currents produced magnetic fields, and that, conversely, moving magnetic fields, or the motion of a conducting material through a stationary magnetic field, produced electrical currents—the flow of energy. As the ability to make precise measurements of these currents, voltages, and fields improved, so did the mathematical description of the relationships between these phenomena. A turning point came in the years 1861 and 1862, when the Scottish physicist James Clerk Maxwell formulated a set of equations encapsulating all these physical relationships, and much more.

At the core of Maxwell's work are four relationships. In the language of calculus, they are partial differential equations. They describe how electrical charge and current relate to magnetic fields and flux in any situation, from a simple charge of static electricity to a complex electromagnet. Maxwell was a brilliant and persistent scientist who published his first scientific paper at the age of fourteen. As he tinkered with his equations, he found they had far broader implications. A magnetic field could typically not exist without an electric field, and vice versa. He realized that this coupling of fields implied that a wave of electrical charge could move—propagate—through a medium together with the complementary wave of a magnetic field. In its simplest form this phenomenon could be visualized as a pair of ropes being whipped into a series of hills and valleys—the shape of a sine wave. When the electric wave reaches a peak or a valley, so does the magnetic wave. The moving electric field produced a moving magnetic field and the moving magnetic field produced a moving electric field. In many senses it resembled a perpetual motion machine. Maxwell also found he could calculate the speed of the motion of this "electromagnetic radiation." To his astonishment, it was the same as the speed of light. Einstein would later write: "Imagine [Maxwell's] feelings when the differential equations he had formulated proved to him that electromagnetic fields spread in the form of polarized waves and with the speed of light!"

Maxwell had discovered, and proved, that light was a manifestation of electric and magnetic fields. It was an electromagnetic phenomenon. This was the final nail in the coffin of Newton's original corpuscular theory of light: electric and magnetic fields had no mass, and therefore light itself was "massless."

Maxwell's equations are still entirely valid today, but for all their beauty and incredible utility, they rest upon something even deeper and more surprising. Different configurations of the electric and magnetic fields do not alter the speed of propagation. Lurking in the equations is the suggestion that the speed of light is constant. There was something else, too. If light was an electromagnetic wave, then surely it needed a medium to move through. Yet light can easily travel through a vacuum. So what was the medium?

Many other physicists took up Maxwell's equations and attempted to explain the propagation of light. The most popular idea put forward by the scientific community was that of a "luminiferous aether," an unseen medium that permeated the universe and allowed electromagnetic waves to get from here to there. But there were problems with this theory. Even if light merrily wiggles its way through an invisible aether at a fixed speed, we should see changes in *apparent* speed. This is because we ourselves move relative to the aether. This could be on foot, on horseback, by train, or by sheer virtue of sitting on a planet that is orbiting the Sun at almost 20 miles a second. The principles of Galilean and Newtonian physics should apply, and the speed of light should appear to vary.

Testing this was an immense challenge. If light travels at 186,000 miles a second, then even the motion of the Earth around the Sun would suggest only a 0.01 percent change in the apparent speed of light in the aether. Measuring the speed of light with some precision in a laboratory is a tricky business even today. In the late 1800s the most cleverly designed experiments and state-of-the-art equipment had fallen far short of the sensitivity needed to detect such a variation between the absolute and apparent speed of light.

Then, in 1887, two American scientists, Albert Michelson and Edward Morley, constructed an ingenious apparatus designed to measure the speed of light with unprecedented precision. Michelson was a well-known optical physicist. He had already expended considerable effort trying to refine the measurement of the speed of light (it was, in fact, a lifelong obsession). He had experimented a few years earlier with a prototype apparatus for achieving a higher level of precision. Now he joined forces with Morley, a professor of chemistry and a skilled experimentalist, to construct the next version.

To avoid even the slightest distortion or vibration during the course of their investigation, they set the apparatus on a massive block of marble that floated on a shallow pool of mercury. This dense fluid supported the weight and let them easily rotate the equipment. For extra caution, the whole thing was assembled in the basement of a particularly solid dormitory building on what is now the Case

Western Reserve University campus in Cleveland, Ohio. To conduct the experiment, a very fine beam of light was split by bouncing it off a partially silvered mirror (not unlike a two-way mirror) at a 45-degree angle, so that two beams were formed at right angles to each other. The beams then traveled to the first of a set of small mirrors placed toward the corners of the marble block. These mirrors reflected the beams back to others across the slab, each carefully aligned to make both beams go back and forth a total of ten times. The final reflection was arranged so that the two perpendicular

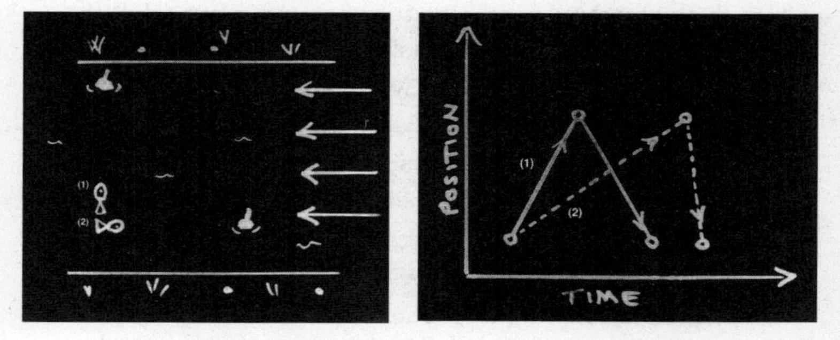

Figure 2. Illustrating the idea behind the Michelson-Morley experiment. Imagine two fish in a flowing river (the aether). They both always swim through their medium by pushing with the same constant force. One fish (1) swims toward and back from a buoy anchored a distance across the river; the other fish (2) swims to a buoy the same distance away, but upstream. Michelson and Morley realized that it takes the fish different amounts of time to complete their round-trip swims, and that photons should behave similarly if they are interacting with a medium. On the right-hand side is a diagram plotting the relative travel position of each fish versus time. The cross-stream fish (1) fights against the sideways current equally on its trips to and from the buoy. The upstream fish (2) must first fight hard as it swims headfirst into the current, but then it speeds back to its starting point. Nonetheless, the cross-stream fish (1) will always manage to arrive back at the starting point first—it appears to have swum faster. This is exactly the principle Michelson and Morley sought to exploit by bouncing rays of light back and forth—upstream and across the stream of the hypothetical aether.

beams would pass through and reflect again on the partially silvered mirror. This time the light would be brought together in one place, inside a small telescope. In this way the light would travel a much bigger total distance, thereby amplifying any variation due to different speeds in the beams.

The Michelson-Morley experiment was brilliantly conceived—in principle. In their travels through the hypothesized luminiferous aether, the beams of light moving back and forth in the same direction as the Earth's orbit would appear to travel at a different speed than the beams traveling in a direction perpendicular to the orbit. The difference in speeds would result in the waves of light from the two beams becoming misaligned. When they rejoined, a phenomenon known as interference would occur; the beams would not mesh exactly. This would be captured in a ghostly series of bright and dark rings that could be measured by the small telescope lined up with the light beams. So, in effect, Michelson and Morley used the very nature of light itself to build the exquisite ruler that they needed to make such a difficult measurement.

It was a beautiful experiment, one that would live on in the history of science forever—because it was a spectacular failure. Within the capacity of the apparatus and Michelson's and Morley's considerable skills, it was clear that the beams of light traveling in different directions had absolutely no discernible difference in speed. This was true regardless of the time of day the measurement was made, the time of year, the position of the marble block, the temperature in Cleveland, or the value of stocks and shares. Either the aether through which the light beams were traveling wasn't operating according to accepted principles of physics or it just didn't exist.

The authors described their experiment in painstaking detail in a journal article in the *American Journal of Science*. Desperate to understand the outcome, they made several proposals as to why they failed to achieve their anticipated result. None sounded too plausible. The only conclusion they could arrive at was that if there really was a luminiferous aether, the Earth couldn't be moving through it very fast.

Later efforts by both Michelson and Morley and others fared no better. All these brilliantly executed experiments failing to detect anything made it devastatingly hard to proceed with the aether hypothesis. Something was afoot.

The second pivotal development that would eventually return Michell's dark stars to the scientific consciousness began inside the brain of a young German patent clerk in Switzerland. Up to this point, the mysterious properties of light had continued to challenge and frustrate physicists—until Albert Einstein published his *special* theory of relativity in 1905. It irrevocably changed our understanding of the nature of reality. In one fell swoop, puzzles such as the fixed speed of light were turned on their heads—suddenly fitting perfectly into place. In fact, Einstein's extraordinary insight came from studying Maxwell's equations. It turned out that they already contained the correct mathematical description of nature. It just required someone to figure out what that was.

There are two fundamental postulates in special relativity. The first is that the laws of physics do not change with your frame of reference, a concept that could be traced back to the Italian astronomer Galileo Galilei. You could be sitting in a deck chair on a tropical island or strapped to a rocket traveling at tens of thousands of miles an hour, yet in either case you would deduce the same laws of physics at play anywhere in the universe around you.

With the second postulate, Einstein went out on an inspired limb. He proposed that the speed of light remains a constant, *independent* of the speed of its source. This is utterly counterintuitive to our everyday experience of the world and the principles of Newtonian mechanics. But it neatly deals with the agonies of Michelson and Morley, does away with the aether, and explains the validity of Maxwell's equations. It also means that the phenomenon of light is an extremely fundamental part of our universe. Today, lasers and more complex experimental arrangements can measure the speed of light with ultra-high precisions of better than approximately 2 parts in 10 trillion. Einstein was right. Light's speed in a vacuum simply doesn't change, irrespective of the motion of its source or observer.

This simple fact has many startling implications for our physical universe. Time itself becomes an important part of any system of coordinates, and the passage of time is relative—it depends on the motion between an observer and any events. The energy carried by moving objects is also modified from the simple classical physics of Newton. Einstein found that even when we see an object as stationary, it still has an energy called its rest mass energy, given by the now famous equation $E=mc^2$. As objects with mass move faster and faster, their apparent total mass, or inertia, increases, approaching infinity if they move as fast as light itself. Einstein reasoned that this meant that no real object with mass could ever reach or exceed the speed of light, since infinite force would be needed to accelerate it to that point.

The special theory of relativity holds in situations in which any relative motion between objects, or between objects and observers, is constant (in other words, where velocities do not change). It was not until a decade later, in 1915, that Einstein published his *general* theory of relativity that fully incorporated modifications for acceleration and the phenomenon of gravity.

If special relativity was a revolution, general relativity was the complete and utter dismemberment of the physics that had gone before it. One of Einstein's critical insights was that if you or I were to float weightless out in the distant, empty universe, it would be *entirely* equivalent to falling in the gravity field of a massive object. This simple observation led him to redefine gravity itself.

The essential point for now is this: general relativity tells us that mass and energy distort the shape of both space and time, curving them as if they were part of a flexible sheet. What we call gravity is really just the way that objects move in this distorted space and time. Even light, which has no mass and a fixed velocity, is subject to its effects. If the path of light is distorted, then light too "feels" the force of this strange phenomenon as its rays are bent around massive objects. Einstein's relativity was one of the most profoundly disturbing ideas of the age. It is still considered a major conceptual challenge, but it provides the best description we have yet of the nature of the universe.

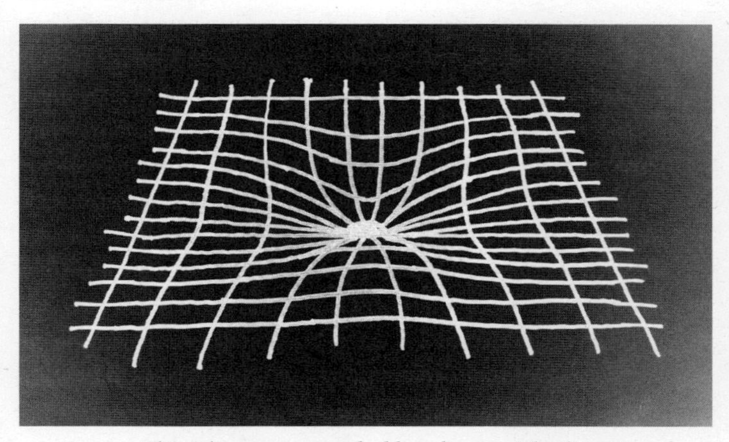

Figure 3. This is known as an *embedding diagram*. The distortion of the geometry of three-dimensional space by mass can be represented as a two-dimensional surface that curves and stretches like a rubber sheet. In this case, an object such as a star or planet is in the middle. Without its mass, the coordinate lines would form a perfectly square grid. With its mass, the geometry of space is distorted—bunched up toward the mass and stretched down. The shortest path through this region may no longer be a straight line. We will explore the reasoning behind general relativity in more depth in chapter 3.

A key consequence led on from Einstein's earlier results. Special relativity had shown that the energy, or wavelength, of light changes with what we measure the velocity of its source to be. A source of light moving toward us will appear bluer, shifted to shorter wavelengths and higher energies. A source moving away will appear redder, shifted to longer wavelengths and lower energies. All the while, the speed of the light stays the same. The size of this effect in our everyday experiences is negligible. Out in the universe, however, objects can move fast enough for these effects to become starkly obvious.

General relativity demonstrated that the same effect occurs in the distorted space and time around massive objects. Light that comes from a source deep inside the distortion around a mass will be seen as shifted to lower energies, or redshifted. The effect is often termed gravitational redshift: photons have lost energy as they have "climbed" out of the gravitational "well" of an object—although their speed remains unchanged. Equally, if an observer is sitting deep inside the distorted space and time around a massive object, then the light arriving from the distant universe will appear to be shifted to higher energies—blueshifted—as it follows its path into the gravity well. Even more disturbing, the distortion of space and time results in events appearing to happen more slowly the closer they are to a large mass, as seen from a distance. Experiments have confirmed this effect. If you had the willpower to sit in a balloon for forty-eight hours roughly six miles above the Earth, you would age by almost 0.0000002 seconds more than someone who had stayed on the ground. Gravity slows time, and this is exactly the same phenomenon as the loss of energy associated with a gravitational redshift.

It took many years after Einstein published his theory of general relativity for some of the implications and details to be worked out. Even Einstein himself did not produce a complete model for how an object like a massive star distorts the fabric of the universe around it. However, hot on the heels of his own breakthroughs, a fellow physicist had an insight that would play a vital role in the application of relativity to this problem.

As unlikely as it seems today, the forty-two-year-old scientist Karl Schwarzschild wrote some of the most impressive works on relativity and quantum physics while stationed at the savage Russian front of the First World War in late 1915. Born to Jewish parents in Germany, Schwarzschild, like Michell, was a polymath with a penchant for astronomy. His genius was recognized in childhood, and by his late twenties he was an established professor in the upper echelons of academia. As war broke out, Schwarzschild dutifully signed up, joining the German artillery. Somehow he continued his scientific work. In a letter to Einstein, he derived a mathematical solution that described the distortion of space and time surrounding a massive spherical object. In a second letter he derived a solution for the curvature of space and time *inside* such a massive object, assuming it was uniformly dense. Tragically, within six months of sending Einstein his calculations, Schwarzschild died of illness on the front, never to see the ultimate implications of his work.

Paramount in Schwarzschild's legacy is a formula that now bears his name. The Schwarzschild radius establishes a relationship between the mass of an object and its effect on light. This was the critical link that would eventually demonstrate that Michell and Laplace's dark stars might actually exist in our universe.

When Michell and Laplace thought about the properties of these massive objects, they mistakenly considered light to be made up of little bodies that would feel the pull of gravity just like a rock or a tennis ball or anything else. According to this theory, we fail to see light emerging from the surface of these stars because it has been pulled back into the star by the force of gravity. But if you could travel toward a dark star, you would begin to encounter these corpuscles of light before they fell back to its surface. If you moved a little closer still, you would see them looping past on their trajectories, like the curving flight of trillions of balls thrown upward and falling back to the ground. All you'd have to do is get close enough, and the light of the dark star would begin to reveal itself.

Remarkably, it is here that Michell's dark stars come crashing into our modern world. In Michell's language, at some distance from a sufficiently massive object the velocity required to escape the gravitational pull of that object begins to exceed the speed of light. Light is halted, and the object is dark to the outside universe. Yet we now know from experiment and fundamental theory that light has no mass, and its velocity remains unchanged. It simply follows the shortest path in time and space. Based on the principles of general relativity, what Schwarzschild's formula suggested was that there is nonetheless a distance from the center of a mass from which light cannot escape further.

Schwarzschild's radius corresponds to a singularity in his mathematical solution to the distortion of space around a spherical mass. A mathematical singularity is simply a point at which an algebraic expression provides no meaningful answer, like calculating the value of one divided by zero. In the case of Schwarzschild's wonderful formulation, such a singularity occurs at a particular distance from a massive object and indicates an extreme curvature of space and time. But, intriguing as it may be, is Schwarzschild's radius just some mathematical tomfoolery, or does it correspond to something observably real? The answer is that while the singularity can be smoothed away by the right choice of mathematical variables, there is nonetheless something remarkable about this location. All paths at this radius turn inward, even for light itself. For you as an outside observer, the light is also redshifted—its wavelength stretched—by an infinite amount. No matter how close you get, you will never see photons coming from within.

Einstein had demonstrated that light is the measuring rod of the cosmos, knit into the very web of the observable universe. It defines the way we experience the world. It defines the way that all matter and energy interact. The Schwarzschild radius is more than a point from which light cannot escape. From the frame of reference of an external observer, it represents the place where time and space seem to come to a halt. If you could place a clock at this location and observe it safely from outside, it would appear to have stopped. Strictly speaking, it would also fade entirely from view, as light coming from it is redshifted to nothingness as it climbs up to you. Anything occurring inside this point, any event, can never be seen in the external universe. For this reason, the Schwarzschild radius is also known as the event horizon.

The most obvious question, and one that came up again and again in the decades following these revelations, was whether such

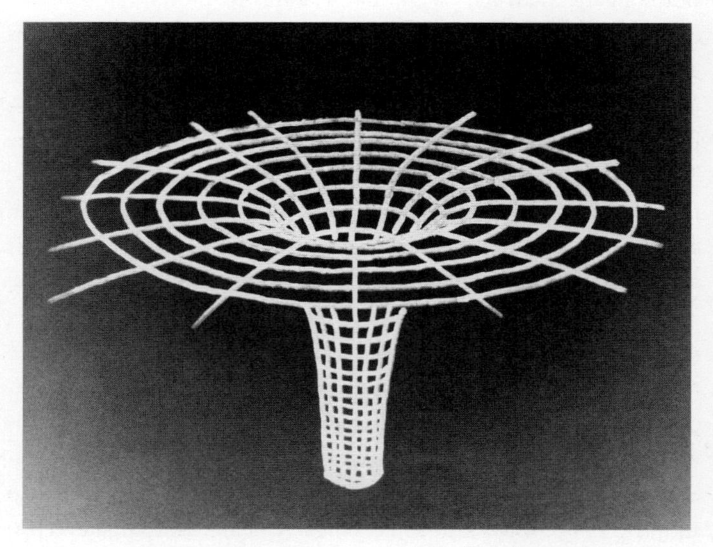

Figure 4. Another representation of the distortion of space around a massive object. In this case, a dense mass is curving space and time to an extreme state. At the bottom of this funnel is the event horizon. The outside universe receives no information from below this point, since even light cannot escape from this far down the funnel.

places could actually exist in the cosmos. The mathematical definition of the Schwarzschild radius is a very simple function, a fixed property of the mass of any spherical object. The tricky issue is that the actual value of this radius is very small. For example, while the Earth has a mass of about 13 trillion trillion pounds (6×10^{24} kilograms), its Schwarzschild radius is only about 9 millimeters—less than half an inch.

Herein lies part of the problem. You would have to pack the entire mass of the Earth within that 9-millimeter radius to create the event horizon. Given the real size of the Earth, there is clearly no point at which space and time ever become distorted enough to prevent the escape of light. Our huge Sun has a mass about 332,000 times greater than the Earth, and a radius of more than 400,000 miles. It would have to be compressed by a factor of more than 200,000 to fit inside its Schwarzschild radius of 3 kilometers, or 1.86 miles. Only then would space and time be distorted enough to prevent the escape of light.

While general relativity had provided a more complete description of the nature of gravity, and a rigorous and satisfying demonstration that dark objects could in principle exist, everyone had a very hard time believing that such nonsensical things could actually be out there.

Ironically, Einstein himself was one of the people arguing against the plausibility of such an object. What Einstein objected to, in company with the mighty English physicist Arthur Eddington and others, was the idea that real places could ever meet the necessary criteria to create an event horizon. There was also no obvious natural process by which any object could be made so compact. This was compounded by the peculiar nature of the event horizon. Time itself would slow to a halt at this point. From the viewpoint of the external universe, this might prevent anything real, with lumps and bumps, from ever vanishing entirely inside this radius. It would be stuck in stasis forever.

There were different ways of framing some of these arguments. Einstein used the example of a cloud of small masses orbiting one another, like stars orbiting in the space and time distortion, or gravity field, of their combined mass. The more compact this cloud becomes, the faster and faster the small masses need to orbit to keep the cloud from succumbing to gravity and collapsing toward its center. If the cloud becomes small enough to shrink within its Schwarzschild radius, then the little objects would have to move faster than the speed of light, which Einstein reasoned was impossible.

Over the next decades, a remarkable cadre of some of the greatest scientists of the twentieth century gradually broke through a series of highly complex and challenging physics problems that would finally resolve this issue. Other extreme environments would turn out to be far more commonplace in the universe than anyone had suspected. These would be the stepping stones to an answer.

› › ›

Beginning in the early 1930s, another revolution in science was well under way. This was the formulation of quantum mechanics, the physics of atomic and subatomic scales and the dual nature of matter as both particle-like and wave-like. If general relativity had toppled our previously tidy picture of the nature of reality off its perch, quantum mechanics took it on an extended bender that few people, if anyone, could or still can completely comprehend.

Many scientists played key roles in the development of this new physics, from Einstein himself to Max Planck, Niels Bohr, Werner Heisenberg, and others. In 1927, Heisenberg was the first to formulate one of the most philosophically challenging and strange parts of quantum mechanics—the uncertainty principle. At the core of this extraordinary description of the physical world was the fact that at microscopic scales nature has an inherent "fuzziness." For example, it is impossible to precisely measure both the location of something and its momentum—its mass multiplied by its velocity. If the location of an object like an electron, which occupies scales on the order of femto-meters (10^{-15} meters), is known to high precision, then its momentum will be very poorly constrained. Because "measurement" always involves actual interaction—for example, trapping the electron in a tiny space—there is no getting around this. This intrinsic uncertainty to the world opens up all manner of deeply disturbing phenomena, from parallel realities to virtual particles, appearing out of nothingness and vanishing again. Nonetheless, seen through the protective shielding of mathematics, quantum theory is clearly a good description of the universe around us. The behavior of atoms and electrons, of atomic nuclei, and of light and electromagnetism is accurately described by quantum mechanics.

As this deeply weird subatomic reality was revealing itself, other critical developments were taking place in stellar astrophysics. It became increasingly clear throughout the early 1900s that stars and star-like objects were not permanent fixtures but constantly evolving. Not only did they come in a wide variety of sizes and colors, but somehow they also represented different stages in the life cycle of a single phenomenon. And the only known energy source capable of powering stars had to be that of nuclear fusion, in which matter is transformed to energy—now described by the special theory of relativity and quantum mechanics.

By the late 1950s the major pieces had fallen into place. We knew that stars were objects in which gravity competes against pressure—the pressure of a mix of electrons and atomic nuclei known as plasma, and even the pressure of light itself. Gravity "tries" to compress, or collapse, matter inward. The outward pressure tries to keep matter from collapsing. This competition results in the cores of stars reaching temperatures of tens of millions of degrees. Such extreme conditions are sufficient for the nuclei of elements to bind or fuse together, forming or synthesizing heavier elements and releasing energy. This is critical, or else life-forms such as ourselves could never exist.

Most of the visible matter in the universe still consists of hydrogen and helium. These are the primordial elemental remains of the hot young universe following what has come to be known as the Big Bang. All the carbon, nitrogen, oxygen, and every other heavy element in the universe came along later. The stars are responsible. By fusing hydrogen and helium into larger and larger atomic nuclei, they act as cosmic pressure cookers, serving up new elements.

The recipes get complicated, but the more massive a star is, the heavier the elements it can eventually synthesize. Also, the greater the mass of a star, the faster it can "burn" up the lighter elements that serve as fuel. While a star like our Sun may cook atomic nuclei for a total of about 10 billion years, a star that is twenty times more massive may eat through its fuel in just a few million years. The least massive stars, a mere tenth of the mass of our Sun, can quietly burn for a trillion years or more.

The ultimate fate of stars was a critical part of these discoveries. A star bereft of its central source of energy is an object in which gravity might win its war with pressure once and for all. This too is a complex problem, but there are signposts in nature. The decades following the early 1900s saw a steady stream of increasingly sophisticated and challenging observations about the universe around us—in particular, the discovery and characterization of distant astrophysical objects that were clearly nothing like our Sun, or its familiar stellar neighbors. Among these were bodies called white dwarf stars. Despite being extremely dim, they exhibited the colors of light that one would expect from a big, very luminous, and very hot star. In the 1920s astronomers realized that these were actually tiny objects, far smaller and far, far denser than typical stars. We now know that their density is such that a mere cubic centimeter, about the size of the tip of your pinky, can have a mass of *millions* of grams. To put that in some kind of perspective, a cube of white-dwarf material only about thirteen feet on each side would have the same mass as all of humanity.

Stellar astrophysics provided an explanation for the origin of such objects as the remnants, the burnt-out husks, of stars like the Sun. However, explaining how such a dense object—although still much larger than its Schwarzschild radius—could exist in a stable state was a much trickier question. For an object as compact as a white dwarf, the normal pressure forces, the same push and shove of atoms that keeps our Sun from imploding because of gravity, simply do not suffice.

The first critical insight into this problem came from the English physicist Ralph Fowler. An athletic and vigorous Cambridge scientist, Fowler had moved hungrily through mathematics to physics and chemistry. In the 1920s he deftly applied the newly minted quantum mechanics to the question. The equations revealed that as matter is forced into denser states, a new type of pressure, with a barely noticeable role in "normal" environments, such as here on the surface of Earth, must come into play. As the atoms in a white dwarf are squeezed together, the electrons are increasingly confined, boosting their momentum and unveiling more and more of their wave-like nature. Quantum mechanics dictates that the little electron waves are not allowed to impinge on each other; the particles must remain distinct. This creates a force known as *degeneracy pressure* that pushes back against gravity in the white dwarf, far exceeding the pressure of a normal gas. Fowler understood that this pressure didn't even depend on temperature. In fact, a white dwarf could, given enough time, cool off to absolute zero and its electron degeneracy pressure could still support it! But was there some limit? How massive could a white dwarf be and still not collapse under its own gravity?

It took the genius of a young physicist training in Madras in southeast India, named Subramanyan Chandrasekhar, to crack the problem, with a piece of insight that effectively married the varied findings of relativity, quantum mechanics, and gravity.

In any stable object with the density of a white dwarf, the electrons end up fizzing about in their tiny compressed volumes exceedingly fast. Speeds well over 50 percent of the speed of light are common. The more massive a white dwarf, the higher this speed gets as the electrons get squashed into less and less space, and their wave-like nature takes over more and more. There are two remarkable consequences. The first is that in contrast to mundane objects like normal stars, the more massive a white dwarf is, the smaller it gets. The second is that since nothing can travel faster than the speed of light, there is a very real limit to how massive the dwarf can be. Eventually the electrons cannot fizz any faster, their degeneracy pressure cannot increase any further, and gravity must overwhelm the object.

Although it would suffer tremendous criticism and take many years to become fully accepted and recognized, in 1935 Chandrasekhar presented his complete theory explaining the behavior of all white dwarfs. He also predicted the maximum mass that they could ever attain. He had realized that this new degeneracy pressure was *only* enough to prevent a white dwarf from collapsing under its own weight if the white dwarf never exceeded a mass about 1.4 times that of our Sun.

There are many other fascinating threads to this tale, but Chandrasekhar's beautiful insight was pivotal. Here were hints at an answer to the puzzlement and distaste felt by Einstein and other physicists over how any real object could come close to inhabiting its Schwarzschild radius. Here too was a linchpin in understanding the life cycles of stars themselves—many of which end up as white dwarfs. It is not surprising that the great modern observatory of X-ray photons, capturing light from 12 billion years across the universe, was affectionately christened "Chandra."

Dissecting white dwarfs was just the beginning. As the nature of stars yielded more and more secrets to human understanding, so did the nature of the subatomic realm. The twentieth century saw an unprecedented entwining of science with the development of weapons and the politics of war and economics. As physicists on both western and eastern sides of the planet raced to build increasingly devastating nuclear bombs, they also pushed forward the science of extreme states of matter. The next piece of the puzzle for dark stars was the realization that an even denser state of matter could exist. Beyond white dwarfs was another possibility, where the electrons were subsumed into the nuclear particles themselves, turning protons into neutrons, to form an object that was in essence a giant and peculiar atomic nucleus—a neutron star. It would be far, far denser and more compact than anything seen before. The American physicist J. Robert Oppenheimer, who had played a central role in the development of the atomic bomb, was one of those who developed the physics necessary to describe such an extraordinary object. Just like the white dwarfs, neutron stars had a limit to their mass. Beyond two or three times the mass of the Sun, gravity would overwhelm them.

Unlike white dwarfs, however, neutron stars had never been observed in nature. This changed in the late 1960s with several intriguing astronomical measurements. The culmination was the spectacular discovery by the scientists Jocelyn Bell and Antony Hewish of a distant neutron star spinning around its axis roughly once a second, assigned to a class of objects subsequently named *pulsars*. The detection of this object came from a giant array of radio antennae, covering about four acres of land amid the fields a couple of miles west of Cambridge in England. Aided by the lawn-mowing skills of a flock of dedicated local sheep, the Belfast-born Bell and her English thesis advisor Hewish were originally planning to study radio emissions from objects in the distant universe. They were shocked when they found this new pulsing signal. As scientists puzzled over the nature of this object, they realized that the only conceivable explanation was a very, very small and very rapidly spinning body sending out a lighthouse-like beam of radiation. The only astrophysical object that could be this small yet tough enough to withstand spinning this fast was the conjectured neutron star.

Neutron stars make white dwarfs look positively tenuous. A mere cubic centimeter—about the size of a sugar cube—of neutron star material has the same mass as all of humanity. While a white dwarf may contain the mass of the Sun within a sphere roughly the size of the Earth or a little larger, a neutron star can contain *twice* the mass of the Sun within its radius of about 7.5 miles.

In a neutron star, gravity is resisted by the same kind of degeneracy pressure as in white dwarfs, except that it is now the neutrons themselves, rather than electrons, providing the force. The incredible compactness of neutron stars brings them much, much closer to being contained inside their Schwarzschild radius. To escape from the surface of one of these objects you would need to move at a substantial fraction of the speed of light—as much as 30 percent, or 62,000 miles a second. Space and time are so distorted or curved that if you fell from an altitude of one meter, you would crash into the surface traveling at roughly 1,200 miles *a second*.

Finally, here were objects in the universe that hovered on the edge of darkness. And together with more detailed and better-understood models of how stellar remains could implode, they provided the final impetus to let go of the cherished belief that nothing could ever really collapse to within its event horizon. If more matter were to be piled onto these bizarre spheres, there is no known pressure force that could prevent utter collapse to within the Schwarzschild radius, and inwards to a single point that is, to all intents and purposes, of infinite density—an inner singularity. By the late 1960s, the reality that such places existed within the cosmos was generally accepted, and observations of the universe were beginning to turn up some intriguing candidates.

In 1967, the American physicist John Wheeler gave a talk at what is now the NASA Goddard Institute for Space Studies at Columbia University in New York. In this nondescript building, which also houses on the ground floor a restaurant immortalized by the singer Suzanne Vega as "Tom's Diner," the charismatic Wheeler used the term "black hole" to characterize an object collapsed within its Schwarzschild radius. It stuck. After a journey of two hundred years, Michell's dark stars had finally become black holes.

Since then we have learned more and more about these extraordinary objects. Earlier on I stated that black holes play a critically important role in making the universe the way it is, and in setting the stage for life itself. That may sound pretty outrageous, but this universe of ours is far more interconnected and far more nuanced than we might have suspected even ten years ago. The concepts that help us get a grip on it are also some of the most important and critical ideas in the physical sciences over the past century or so. We've encountered a few of them above. They include the finite and unchanging speed of light; the nature of space and time, mass and gravity; and, of course, the finite age and scale of the observable universe. I've touched upon others: the nature of stars, the quantum universe, and the synthesis of elements from primordial hydrogen and helium. Beyond those are further components, ideas that are still at the very cutting edge of human understanding: the ways in which this universe makes stars in the first place; the formation of worlds; the molecular structures that pervade interstellar space yet are the same flavor as those that make life on a planet. It's a remarkably diverse set of ideas, and so some clear perspective would be a good thing.

We have already traversed the cosmos from a colossal black hole in a now-ancient galaxy to our own microscopic speck of rock and metal. But what do we know about the size and shape of the observable universe? What does it look, feel, and smell like? If we are to understand what forms it, what makes it appear the way it does, and how to navigate its highways and byways, peaks and plains, nooks and crannies, we will need to begin with a very, very good map.

2

A MAP OF FOREVER

Making a map of nature is such an instinctive and appealing notion. As a species, we have found it incredibly difficult to resist mapping and charting everything we see around us. Not only do we like to know where we are at any given moment, we also like to know the arrangement of the world far beyond our visible horizon. Maps help us organize our model of the world, and they can help us test whether that model fits reality.

For more than ten thousand years we have scraped, chiseled, drawn, and painted charts of our immediate surroundings and of the larger regions beyond. Driven to explore, we have expanded the boundaries of the known world from our every point of origin. What was once a blank expanse labeled with helpful information like "Here Be Dragons," we have transformed into familiar and well-known terrain. Over time, though, terrestrial maps have reached their limits—the inevitable consequence of living on a sphere. Bereft of new continents, eager cartographers have looked at finer and finer levels of detail, filling in ever more minutiae. Today, many of us can sit in front of a computer and zoom instantly to whatever distant corner of the globe our whimsy takes us to. We can even descend to a street's-eye view of places we may never travel to in person. Just like all the humans before us, we may long in vain for a still closer look: a peek inside that shop in an unknown town, a glimpse of the headlines on the newsstand in a country we'll never visit.

Charting the objects we see in the sky has both paralleled and diverged from terrestrial mapmaking. Astronomy is the most ancient of sciences, and it has always been about mapping, and relating to, the shapes and rhythms of nature. The wonderful propensity of the human brain for pattern recognition has enabled us to imagine great richness in the night sky as well. A grouping of stars like the constellation we call Orion is linked to several different myths in cultures throughout the world. The Australian aborigines looked at Orion and saw a canoe carrying two banished brothers. The Finns saw a scythe. In India, that pattern of stars was obviously a deer. For the Babylonians it was the heavenly shepherd, and for the Greeks it was the hunter, a primordial giant.

Astronomy has long motivated both precision and imaginative abstraction in a multitude of societies and civilizations. We find in the recorded histories of practically every known culture that people have developed methods to plot out the locations of stars and planets. Indeed, from Oceania to Asia, from the Middle East and Africa to Europe, and in the Americas, we have keenly observed and depicted the night skies. Also, archaeologists think that Orion is one of the oldest recognizable celestial representations, carved into the ivory of a thirty-three-thousand-year-old mammoth tusk in Germany; another is the Pleiades star cluster in the paintings of the Lascaux caves in France, made some twenty thousand years ago. In the southern planetary hemisphere the indigenous peoples of Australia have complex pictorial and oral depictions of the stars that are a part of their forty-thousand-year-old cultural history.

The curved surface of the Earth has prevented us from easily mapping the globe. But the skies have always been fully accessible to anyone with an interest in making sense of our place in the cosmos, or simply dreaming about what magical realms might exist elsewhere. However, our eyes are limited in their sensitivity. Fainter celestial objects become invisible, and close-knit points of light become impossibly tangled, often merging into confusing blobs of luminosity. A vast and three-dimensional universe is projected onto our retinas as a flat image. What is far may seem near, and what is near may seem far. Mapping the cosmos has therefore been a gradual progression, both outward and across, locating the brighter objects and then filling in the gaps. We have helped our eyes along by constructing telescopes that can capture far more light than our little biological lenses can, and that bring that light to a much sharper and finer focus. These instruments have also allowed us to overcome one of our greatest scientific handicaps. Evolution has equipped us with remarkable senses, yet has blinded us to most of the universe by giving us eyes that detect only a narrow range of photon wavelengths. Eventually, though, we figured out how to build telescopes that reveal much more than just visible light, opening up the whole electromagnetic spectrum and illuminating phenomena beyond our wildest imaginations.

Seeing the cosmos for what it is has required us to overcome many other blind spots, including a perpetually sticky one for mapmaking: we place ourselves at the center. It is a natural assumption and often a practical necessity, but it has also been an enormous hindrance to developing an accurate model of the universe. It took the insight and intellectual conviction of Galileo and Copernicus to challenge the orthodoxy that our home the Earth, like us, was at the center of the cosmos. Relinquishing that child's-eye view was no small accomplishment, but the notion that our whole solar system was nonetheless located somewhere at the center of the visible universe was still very much in vogue even into the first decades of the twentieth century. As so often in science, only a major advance in the observation of nature would convince the world otherwise.

› › ›

Harlow Shapley was born in 1885 in rural Missouri, along with his twin brother, Horace. Bright and possessing an impressive personal energy that would be a hallmark throughout his life, Harlow still faced the challenges of his times. His early education was rife with obstacles. As a small child he was placed in a one-room schoolhouse where his older sister sometimes acted as teacher. By twelve he had dropped out. Then, after studying at home and working for a newspaper, he returned, newly motivated, and completed his schooling in record time. His aim was to get into college and study to become a journalist. But on arriving at the University of Missouri, he discovered that the journalism department wouldn't open for another year. Rather than leave, he decided to sign up for astronomy because it was close to the top of the alphabetical list of subjects, and, he later claimed, because it was easier to pronounce than archaeology. On graduating from Missouri, he was awarded a fellowship at Princeton. Remarkably, just a few years later, in 1914, he had obtained his Ph.D. in astronomy.

The freshly minted Dr. Shapley was well versed in many of the latest astronomical techniques. One of these made use of the properties of a peculiar class of old stars called Cepheids, named after the star Delta Cephei. These objects vary in brightness over days to months on a regular basis as their outer layers undergo cycles of heating and cooling. Surprisingly, the timing of the variations reveals the true luminosity, or energy output, of these stars—the slower the variation, the more luminous the Cepheid. Other classes of stars, like the fainter and rather unpoetically named RR Lyrae variables, exhibit very similar properties. Knowing the true brightness of these variable stars allowed astronomers to deduce their actual distance from the Earth, simply by measuring how faint they appeared. It was like having a marvelous cosmic yardstick, and Shapley knew just how to put it to use.

In the early 1900s it was the general consensus that the Sun was somewhere near the center of the universe. This universe consisted of the great wheel-like disk of stars of the Milky Way, along with near and far nebulae and dense spherical groupings of stars known as globular clusters. It was not yet understood that most of the small hazy nebulae seen in the sky were actually other distant galaxies. Something rather odd, however, had attracted Shapley's attention. The globular clusters, extraordinary orbs of a hundred thousand stars packed close together, were distributed disproportionately across the night sky—most of them were in only one hemisphere. This was extremely puzzling to him. If we stood near the center of everything, then why were the globular clusters off to one side? Shapley, moving on from the halls of Princeton and freshly installed at the Mount Wilson Observatory high in the San Gabriel Mountains above Pasadena, California, set to work to measure the distances to these stellar hives.

His measuring tools were the variable stars they contained. He was determined to map out the three-dimensional locations of the Milky Way's globular clusters. By 1918, after studying a total of sixty-nine of these stellar baubles, he had his answer. The globular clusters were asymmetrically distributed on the sky because we were not at their common center. Like geographical markers, they plot out a great sphere centered not on us, but on the Milky Way itself. This meant that our solar system was not at the center of the galaxy, it was stuck out in the suburbs. As Shapley himself put it: "The Sun is very eccentrically situated in the general system . . . the local group is about half way from the center to the edge of the galaxy."

It was an extraordinary discovery. Surprisingly, it was met with general agreement by other astronomers—it just made sense. But it opened the floodgates for what were far more controversial revelations in the following decades. We would find not only that the Milky Way is merely one of many galaxies, but that there is no center for these objects themselves. All of them are flying apart as the universe expands. Shapley's great leaps in 1918 were the beginnings of the modern mapping of the cosmos.

But challenges abound. As if it isn't enough that beyond the confines of our spherical world the universe seems to stretch on endlessly, even the plainest of places can hide the most extraordinary complexity.

There is a famous issue in mapmaking that illustrates this particular challenge. It's very much an earthbound problem, but its implications go far beyond. Awareness of this conundrum originated in the 1950s, in the work of an English scientist by the name of Lewis Fry Richardson. Richardson had eclectic tastes in his research, from physics and meteorology to the mathematics of war. As a Quaker born into a well-to-do family in 1881, he believed strongly that morality was paramount, and his experiences as an ambulance driver in France during World War I only reinforced his pacifism. Viewing war as a disease and an affliction, he eventually tried to model armed conflicts with mathematics. In his book *Arms and Insecurity*, published in 1949, he famously stated: "The equations are merely a description of what people would do if they did not stop and think."

These studies led Richardson to wonder whether the length of a shared border between two countries was somehow related to the probability that they would go to war. As archaic as this hypothesis may sound in today's world of transglobal aggression, it nonetheless fit into a much more complex statistical analysis with real bearing on the nature of human conflicts.

To test the predictive power of his sophisticated mathematical models, Richardson needed to research the estimated lengths of real country-to-country borders. To his great surprise, the numbers he found differed wildly depending on his reference source. The lengths of well-established, settled borders, such as those between Spain and Portugal or the Netherlands and Belgium, could differ by hundreds of miles from one map or survey to the next. What could be going on?

Richardson discovered that the answer was related to the smallest unit of measurement used to estimate a border's length. If a surveyor went along the border and made visual measurements with the telescope of a theodolite every few hundred feet, adding up each distance, he would get a very different answer than if he had trudged every foot with a surveyor's wheel. The problem, as Richardson eventually surmised, is that natural borders or coastlines have effectively infinite complexity. The smaller your measuring rod, the more of its lengths you can fit in around every kink and turn—and the longer the border appears. Two runners, one with a short stride and one with a long stride, literally have to travel different distances if they race along the very edge of an island, or a country's border.

Richardson, with his rather morbid but morally driven fascination with war, had stumbled upon one of the key elements of what would later become known as fractal mathematics. His careful analysis of the discovery became a vital touchstone for this field. Many things we see in nature are best described not by straight lines or simple curves, but by endlessly repeating patterns or structures nesting inside one another.

On the face of it, the universe around us may not seem quite this complex. Indeed, there are very distinct objects and structures that appear superficially to be finite and straightforward. Stars are stars, galaxies are collections of stars, and galaxies themselves form greater collections known as clusters and even superclusters. We can readily make a chart showing the positions of bright stars in the sky or the locations of visible galaxies. But this is deceptive. Just as the earliest maps of the world blithely cast entire continents as "desert" or "land of giants," sweeping cartographical judgments of the sky are invariably incomplete. What may appear to be a single star can very often turn out to be a pair or even a triplet of objects, hiding inside imperfect images made by our eyes or by our telescopes. A galaxy is composed of stars, but it is also composed of a gas of discrete atoms and molecules, forming great interstellar clouds that can be thin and almost invisible, or dense and bulky like the great nebulae in Orion and Carina. Galaxies can also contain vast quantities of microscopic dust composed of silicates and carbon.

There may be planets lurking around at least 50 percent of all stars in a galaxy like ours—cold, dark objects completely lost to view against the glare of their suns. And what about those planets? Don't they in turn deserve mapping? Don't they have geographical features that descend into ever smaller levels of detail? Of course, here we are utterly limited. Beyond the worlds of our own solar system we have for now only the most rudimentary information about planets around other stars—the exoplanets. A telescope big enough and sensitive enough to genuinely map the surface of any of these distant worlds has yet to be made, but there's every reason to believe that eventually it will be.

Clearly, the observable universe shares some of the intractable difficulties we encounter in trying to evaluate the lengths of country borders or coastlines. Complex objects are nested within complex objects. But this is not the only huge challenge for cartography of the cosmos. The great majority of the information we have traditionally sought for maps of the Earth comes from the observation of visible light. Whether by peering through a telescope from the deck of a ship warily sailing offshore or by trudging through the interior of some great continent, we make a map of what we can see and what we can measure by seeing. In many respects, this allows us to capture the information about the Earth that is most useful to us.

Out in the universe, however, an abundance of environments and physical phenomena emit or absorb light at wavelengths our eyes simply cannot respond to. So another critical aspect to mapping the universe is to incorporate not just the colors and spectra of objects as we might perceive them, but to include the vastness of the entire electromagnetic spectrum. This extends from long radio waves through microwaves and far and near infrared to visible light. It carries on to ultraviolet light, to soft and hard X-ray photons, and to gamma rays. And there are even more exotic aspects of matter and energy to probe, from gravitational waves and neutrinos to fast-moving subatomic particles, not to mention the pervasive but still mysterious dark matter, whose presence reveals itself only through the gravitational pull of its mass.

So what does our current map of the known universe look like? It already contains a vast amount of information, yet it is barely a scrap of parchment compared to what would be the full atlas. By physical necessity, the level of detail in this map drops off as we move farther and farther beyond Earth. The map also contains many, many layers of data from the different photons used in its construction, each revealing a different but related topography. Often these multiple layers exist in tiny patches, the results of someone's hard work with a telescope on a very particular object or set of objects and no others. Small pinpricks on the sky where the density of information is great are surrounded by areas of the map that are still somewhat blurry and empty, and may contain dragons.

Doing this map justice in words or images is extremely difficult; it already holds so much information, even if it is only a tiny fraction of what there is to know. It is both three-dimensional and four-dimensional, as time becomes linked to what we see. The farther away objects are, the longer their light has taken to reach us, all the way back through the 14-billion-year history of the universe. There are so many categories of objects and phenomena, and so much higgledy-piggledy data from several hundred years of telescopic astronomy. The best we can do to gain some amount of intuition for this atlas is to play out a thought experiment, a parlor game to let us begin to grasp what a map of forever would look like.

›››

Let us pretend that a very large box has just been delivered to our doorstep and we have hauled it inside. Within the box is an ominous-looking sack filled to bursting with a mysterious shape. The occasional wisp of gas trickles out through the knotted top, and every so often a muffled thump and an obscured glow come from within.

This sack contains what we could regard as a representative portion of the universe. Cosmologists often speak in terms of "fair samples" of the universe. What they mean is a large chunk, or volume, of the stuff Out There. It encompasses enough of the varying lumps and bumps, galaxies, galaxy clusters, empty places, and crowded places to be typical. If you surveyed all the properties inside the volume and took an average, it would be very close to the universal average. For example, if you divided the total mass in the sack by its volume, you would obtain a very good estimate of the average density of the universe as a whole. Equally, if you measured just how lumpy the arrangement of galaxies was within this volume, the answer would be a close match to the universal "lumpiness" of structure.

So let's take a peek inside the sack and begin to add up what we find. The first thing that happens when you cautiously peel apart the opening is that electromagnetic radiation floods out, together with particles—such as neutrinos and fast-moving protons, atomic nuclei, and electrons—that are born in highly energetic and violent physical environments. There is also a gentler seepage of hydrogen and helium nuclei and atoms, together with something hard to spot: an as yet unknown species of heavy but invisible particle, known as dark matter. All these things, photons and particles, permeate the cosmos and are important to understand.

Photons of light are present in huge quantities. They come in all flavors, from extremely low-frequency radio waves, where a single crest-to-crest distance may span kilometers, to microwaves, infrared, visible, and ultraviolet frequencies, and on to the realm of X-rays and gamma rays. One of the most pervasive types of photons is the kind that originated in the very young universe, when it was extremely dense, hotter than 5,000 degrees Fahrenheit, and temporarily opaque. By 380,000 years after the Big Bang, the expanding cosmos cooled down and thinned out enough for these photons to fly free, skirting the atoms of hydrogen and helium that would otherwise trap and scatter them. They now fill the cosmos, their wavelengths stretched by the expansion of space and time itself following the Big Bang. This has lowered their energy, and today they mostly span radio wavelengths, ranging from the frequencies where cell phones, TVs, and microwave ovens operate to shorter waves, or higher frequencies, entering the beginnings of the infrared spectrum. They are known as *cosmic microwave background* photons. There are roughly 410 of them per cubic centimeter of the local universe at any single instant. That may not sound like much, but our sack—our fair sample of the cosmos—is hundreds of millions of light-years across and will contain a colossal number. Even the volume of a sphere that encompasses our tiny little solar system, whose very outermost extent may be about one light-year from the Sun, contains at any given instant roughly 10^{57} of these ancient photons. That is more than a trillion trillion trillion trillion. Adding to this impressive count are all the other photons that have originated from stars and cooling gas out in the universe. Light may not have mass, but the cosmos is thick with it. This is an extremely important component of the universe. We will find that it can play a critical role in a number of processes. Space may appear largely empty to us, but in reality it is a heaving soup of unseen photons racing back and forth across eternity.

The other particles that come pouring out of the sack are more difficult to put definite numbers to. Neutrinos are extremely low-mass subatomic particles, less than about a millionth of the mass of electrons. They play a key role in what are termed weak interactions in physics, and they come in a variety of flavors: electron, muon, and tau. For example, one type of natural radioactivity occurs when a proton in the nucleus of an atom turns into a neutron through a process called beta decay. In this transformation, the atomic nucleus spits out a high-speed anti-electron together with an electron-neutrino.

Neutrinos have been likened to the "ghosts of the cosmos," since they have very little to do with normal matter, passing through gases, liquids, and solids with minimal likelihood of actually hitting or interacting with anything. Deep inside normal stars, the processes of nuclear fusion make neutrinos in abundance. But to a neutrino the universe is almost completely transparent, and so they immediately escape from stellar cores and stream out into space. Here on Earth, every second roughly 65 billion neutrinos from the Sun's core pass through every square centimeter of your skin. Eight minutes ago they were produced in the solar center, and they have raced outward at a rate that comes close to the speed of light. Despite that incredible barrage, the chances of one actually being stopped by your body are so low that it may happen only once or twice during your lifetime. Stars are therefore actively flooding the universe with neutrinos. In addition to these fresh ones, there are ancient neutrinos. These are the remains of a stage in the universe's earliest evolution, about two seconds after the Big Bang. Because they are lower in energy, these neutrinos have not yet been conclusively detected, but we expect them to be streaking through the cosmos in all directions.

The majority of normal, recognizable matter that emerges from our open sack is in the form of hydrogen and helium, in the proportion of roughly seven hydrogen atoms to every one helium atom. Again, these are remnants of the young, hot universe. Anything else in this mix of normal matter is a tiny trace by comparison, and has been forged inside stars. The next most abundant element we find in our sack of universe is oxygen, and there is only one oxygen atom for roughly every 1,500 hydrogen atoms. All the elements that are so critical in making objects like planets, and the molecules that are part of us and all living things, are rare—quite literally cosmic pollutants.

Some of these elements come zooming out of the sack with considerable speed. In this case they are components of hot gases, often so hot that most of the electrons that usually stick to an atomic nucleus have been stripped away, leaving an electrically positive object known as an ion. A gas in this state is also referred to as plasma, and what escapes from the sack can have a temperature of tens of millions of degrees. Other normal matter seeps out at an extremely slow rate. These are components of much, much colder gases, some barely a few degrees above absolute zero.

Within this colder gas are molecules. Most of the molecules, like most of the single atoms, are hydrogen. Two electrically positive protons, bound together by their electromagnetic lust for two negatively charged electrons, form a hydrogen molecule. There are also tiny traces of more complex structures: compounds such as carbon monoxide, a carbon and an oxygen atom also bound together by quantum forces; carbon dioxide, with two oxygen atoms; H_2O; and even alcohols. If we sniff very carefully at these whispers of heavy molecular structures, as rare as their constituents are compared to hydrogen and helium, we find that most of them contain the element carbon. In fact, roughly 70 percent of all the heavier molecules that are adrift out in the universe contain carbon. We call these organic compounds. This is surprising, because many of these carbon-based molecules are the same structures that we find here on Earth. What they are doing out in interstellar space is a question that we will visit again later, since it has far-reaching consequences.

All these ions, atoms, and molecules drift out of our sack in an extremely thin haze, tenuous enough to be mistaken for a vacuum. In the emptiest regions of the universe there are just a few hydrogen atoms or molecules per cubic meter. Even when our sack opening lets out some of the denser stuff from places like the rich nebulae, the density only reaches about a trillion atoms per cubic meter. By comparison, the air that we breathe on the surface of Earth contains well over a trillion times *more* particles, roughly 10^{25} atoms or molecules per cubic meter. Our whiff of universe from the sack is gossamer thin, like the finest and most exquisitely subtle perfume.

Among the normal matter, the light, and the neutrinos emerging from this piece of universe is something else that we can barely sense. It neither reflects nor absorbs electromagnetic radiation. The only sign of its presence is the gravitational pull of its mass. This is the enigmatic substance we call dark matter. Exactly what it consists of is still a puzzle. The most likely candidate is a variety of subatomic particle that has very weak, very ineffectual interactions with "normal" matter. Much like the neutrino, it can drift right through solid material as if it were nothing more than a thin fog. Unlike the neutrino, though, dark matter moves slowly, and each particle carries a significant mass. Until we take a proper look inside the sack, we won't see how this mysterious stuff arranges itself within the universe, but by counting what seeps out we can already tell that altogether it outweighs all normal matter by a factor of five. The mass of the universe is dominated by it.

There is one more thing that sneaks out of our sack as we carefully peel it open. Especially in the colder, denser, more sluggish wisps that emerge, for every hundred or so atoms or molecules, we will count one microscopic particle or grain of dust. This is not the same kind of dust that you find under your bed. This is far finer and very different in composition. A typical grain of interstellar dust is only about 0.001 millimeters (one micron) across, and may be composed of compounds such as silicon carbide or graphite. Some grains are even smaller, composed of just a few hundred atoms—barely more than giant molecules.

Where does dust come from in the depths of the universe? In truth, astronomers are still trying to understand the details of its origins, although at least two environments are known to produce these microscopic structures. One is the expanding refuse from big, old stars that are beginning to shut down the processes of nuclear fusion in their cores. As their interior undergoes these changes, their outer parts become bloated and end up being blown outward by the pressure of light itself. As this element-rich gas expands away from the star, it cools off, and just like water condensing out of steam, some of the carbon and silicon and other elements condense into small, solid particles of dust. These are set adrift into the cosmos. The other place where we know dust is made is in the great clouds of material blasted free by the supernova explosions of old stars.

All this seeping and rushing of particles and material occurs before we even look properly into the sack. A lot of stuff in the universe is small and fast-moving or small and drifting. Much of it is effectively invisible to us, either because it doesn't interact with light, or because, like light itself, we can only sense it when it actually collides with us. Yet if you could see the paths of all these components from afar, the cosmos might appear as an opaque fog.

Now let us peel away the sack and admire the structures within. We've analyzed the major ingredients, but how are they arranged, and what are they doing? The first thing we notice is that space is filled with, well, space. Apart from the flood of photons and neutrinos pervading the universe, it is mostly empty. This is perhaps not surprising, but just how sparsely matter is distributed is not immediately obvious to you or me here on Earth. In particular, the scale of atoms and particles compared to the voids between them is tiny. This is strikingly different from our day-to-day sensory experience of life on our planet. But then again, our senses are limited. We think our bodies—and our houses, villages, and cities—take up significant volumes compared to their immediate surroundings. We can reach across the table for a cookie or put our fingers around a pint of beer looking lonely on the other side of a bar. This apparently high density of matter is all a question of perspective, though.

A hydrogen atom consists of an electrically positive proton and an electrically negative electron. Allowing for the quantum fuzziness of matter on these scales, the proton is roughly a thousand-trillionth, or 10^{-15}, of a meter across. The electron can be thought of as occupying a scale a thousand times smaller than this. Yet the overall size of a hydrogen atom is a ten-billionth, 10^{-10}, of a meter. So the typical space between the proton and the electron is a hundred thousand times the size of the proton. You and I, composed of atoms, are mostly emptiness. This remarkable discrepancy between object size and the space in between extends to much larger scales, too. The distances from the Sun of the planets in our solar system are tens of thousands of times greater than the size of the planets themselves.

As we peer at the distribution of gas in our representative chunk of the universe, there is an average of about one atom of normal matter per cubic centimeter. This means that in any random location an extraordinary gulf typically exists between particles of matter out in the universe. You would have to separate grains of sand by about sixty miles to make an equivalent sparseness. Yet if we step back far enough, clear structures and forms emerge.

We know that our limited human visual organs pick out a very narrow range of wavelengths wherever photons are being generated and reflected. Over the volume of what is contained in the sack, we nonetheless see a glistening arrangement of light. Entire galaxies appear as tiny hazy smudges scattered throughout this three-dimensional space. Their positioning is both random and structured. It is as if the great abstract expressionist artist Jackson Pollock were given carte blanche to paint the universe. Vast gatherings of galaxies emerge as you squint your eyes. In some places there are thousands clustered together within just a few tens of millions of light-years: great cathedrals of light, with galaxies forming vast spherical clouds that increase in density toward the center pulpit. Leading into these glowing monuments are what appear to be strands and sheets. Sometimes they are the barest outlines sketched by the distribution of the small patches of light from galaxies, sometimes bold chains of brightness with galaxy after galaxy tracing out the structure.

Then there are huge zones of emptiness, voids like great open soap bubbles bumping up against one another. As in a painting of an optical illusion, the entire contents of the sack can be seen in two ways. Either it is a clump of these dark bubbles, outlined by the light of galaxies, or it is a network of interconnected threads, sheets, and clusters of light surrounded by darkness—the dendrites and neurons of some megalomaniac artist's impression of a web-like cosmic brain.

Figure 5. A chart of 1.6 million galaxies surrounding our location in the universe. Individual galaxies detected in infrared light are shown as tiny bright, grainy points in a map projection of the entire night sky. The dusty plane of our Milky Way galaxy runs around the edges of this picture and down the middle as a dark streak. The galaxies trace out the web-like and foamy distribution of matter. The bright clump at the center of the map is named the Shapley Supercluster in honor of Harlow Shapley. It is 650 million light-years away and contains more than twenty clusters of galaxies, each containing hundreds to thousands of galaxies.

This is the essence of the large-scale structure of the universe. It is both frothy and stringy, yet mostly empty. Galaxies may be hundreds of thousands of light-years across, but only in their densest clusterings are the spaces between them comparable to their size. The vast bubble-like voids can extend over 100 million light-years, with barely a galaxy within. If we look more carefully, seeking out the atoms and molecules of hydrogen and helium that we know occupy the universe along with the particles of ghostly dark matter, we find that they, too, trace the same structure, the same arrangement that is betrayed by the bright galaxies.

Now we take out our magnifying glass and peer more closely at these hazy points of light, the galaxies. Most of the visible photons coming from them originate from stars. Sometimes it is reflected, or absorbed and then re-emitted from gas and dust that coexist with the stars, but mostly it comes from the incredibly compact stellar objects that make up the galaxies. And the littlest of the galaxies are dwarfs compared to the biggest, which can be a hundred times greater in diameter. The dwarfs may contain only 100 million discernible stars, compared to the hundreds of billions in a giant galaxy. The stars themselves, across all galaxies, are highly varied. The smallest red and dim objects are barely a tenth the mass of our Sun and twice the diameter of the planet Jupiter, gently trickling out a thousandth of the energy that is produced in our own solar system. The brightest and bluest objects are over ten times more massive than our Sun and have ten to twenty times its girth. They pour out hundreds of thousands of times as much energy, but are far, far rarer than Sun-like stars. Most of the stars throughout all the galaxies in our sack of universe are less than half the mass of our Sun. Then the very rarest of all normal stars are those that are a hundred or so times the mass of our Sun, but their numbers add up to barely a grain or two among the great dunes of stellar objects.

All these very different stars, though, are in the midst of their normal adulthood. Deep inside their cores primordial hydrogen is being fused into heavier elements. For the smallest stars, this is a very, very lengthy process. Our Sun will perform this task for a total of approximately 10 billion years. Stars a tenth of the solar mass are the slow cookers of the universe, taking at least a trillion years before they deplete their primordial nuclear fuel. The most massive stars are gluttons. Super hot in their bellies and gulping through their food, they will consume hydrogen for perhaps only a few million years.

Many more star-like objects also exist in the galaxies inside our sack, but these represent the very young and the very old. At one extreme of the stellar life cycle are protostars, objects not yet ready to fuse elements; at the other extreme are the remains of stars that have lived and died. White dwarfs are cooling lumps of dense, leftover stellar material. Held up by the quantum pressure of electrons that we encountered earlier, they are almost as numerous as stars that are still generating energy. Neutron stars are similarly common, since they are the descendants of the most massive stars. These blazing objects are rare in their prime, but they live fast and die young, littering a galaxy like the Milky Way with their brutally compact and dense corpses.

Then there are the black holes. In a big galaxy, small black holes, a few times the mass of our Sun, may number in the thousands or even tens of thousands. Sometimes matter falls screaming into their gravitational lairs, releasing its energy as brilliant flares of electromagnetic radiation. Much more rare, but much more cosmically important, are the giant black holes—the lords of gravity. Billions of times more massive than our Sun, they sit imperiously within the deep recesses of galaxies, a great mystery for us to explore.

Our own Milky Way is a big galaxy, among the largest of such objects. Galaxies like it contain several hundred billion stars, protostars, and stellar remains. It sounds like an awfully packed environment. While it can become so toward the center, a galaxy, much like the rest of the universe, is still mostly empty space. A star the size of the Sun is about 30 million times smaller than the space between it and the nearest star. Even in the densest regions of our galaxy, the average separation between stars is still about a hundred thousand times their sizes. Mapped by its stellar components, a galaxy is a remarkably *open* place.

Just like the stars they consist of, the galaxies we find inside our sack of universe have an array of forms and types. In addition to coming in different sizes, they exhibit starkly different outward features. The most noticeable is that galaxies range across two primary structural forms. Some are great flattened, disklike structures with huge curving rivers of stars, as if paint had been casually dribbled onto a spinning, slippery plate. Others are almost spherical, dandelion-like hazes of stars. Those that are great disks amount to perhaps 15 percent of all the galaxies we see, and are fittingly known as spirals, of which our Milky Way is one. The fog-like clouds of stars, some of which are flattened into great ovoids, are known as ellipticals. Because this form includes many of the smallest galaxies, the ellipticals outnumber the spirals several times over. In between these major classes are all manner of hybrids, as well as galaxies that have clearly suffered gravitational trauma. Lumpy, distorted, and shredded, these are collectively known as "irregulars."

The largest galaxies take the form of giant ellipticals. These cloud-like structures get denser and denser toward their centers, with more and more stars packed ever closer together. Most often these mammoth stellar collections sit at the middle of the great clusters of galaxies, at the apexes of the cosmic webbing that we first noticed. In stark contrast, the spiral galaxies typically avoid these locations, preferring to lurk out in the suburbs and hinterlands of the universe. Unlike the pure ellipticals, the great spiral disks slowly spin, although their swirls of stars move separately from their mass, like a projection of light onto the platter of a turntable. The two types of galaxies also show markedly different stellar colors. Spiral galaxies often contain many young, bright, hot, massive blue stars along with thick gas and dust, while elliptical galaxies are usually composed of older, smaller, red stellar systems, with few of the beautiful nebulae that spirals contain. Whatever has happened to make these great collections of stars so different is deeply rooted in their pasts, their locations, and their gravitational environments.

Within all galaxies, no single star is motionless. In the ellipticals and dwarf galaxies, most stars are flying to and fro, on orbits that are little more than a single track that takes them in close to the galactic center and then out again on the other side. They are like angry hornets flying back and forth across the central hive. In a spiral, the stars out in the great disk follow circular orbits around the center of the galaxy, wobbling in and out and up and down as the lumpy nature of the system pushes and pulls on them. Our own Sun does this in our galaxy, completing a leisurely circuit every 210 million years. Toward the bulging centers of spirals, though, stars begin to buzz around more like those in elliptical galaxies.

All galaxies are themselves in constant motion. Within the great clusters, galaxies behave much as the stars do in elliptical galaxies—flying in and out of the core. Velocities can be huge. A galaxy picks up a lot of speed as it falls toward the center of a cluster's gravity well, reaching more than six hundred miles a second in many cases. Out in the intergalactic countryside the pace is calmer. Average velocities are only one-third as fast. Yet here too the great sculptural forms of threads, sheets, clusters, and voids prevail. Matter tends to stream along the web-like threads, and around the voids. It is gradually filling the densest, most massive clusters and superclusters with more and more material, the way gullies feed mountain lakes.

Putting aside our magnifying glass, we pick up a powerful microscope to pry ever further into this veritable cosmic zoo of galaxies and stars in our sack. Here again we meet up with Richardson's intractable mapmaking problem: the endless complexity of borders and coastlines. Within every nebulous cloud of molecular and atomic gas are an infinity of corridors and surfaces. Orbiting more than 50 percent of all stars anywhere are big and small planets. Some have moons and satellites, which in turn have smooth or ragged surfaces, mountains, valleys, and even intricate, endless coastlines. Then there are shadowy components: objects called brown dwarfs that are neither big enough to be stars nor small enough to be comfortably called planets. As we tune in to the further reaches of the electromagnetic spectrum, more and more extraordinary structures light up, coming magically into view. Great radio-emitting loops of magnetically entrained particles extend from stars, and even from entire galaxies. High-energy photons, from the ultraviolet to the X-ray to the most energetic gamma rays, stream from the surfaces of fiercely hot white dwarfs and neutron stars. Ten-million-degree gas trapped in the gravity wells of galaxies and galaxy clusters glows with X-ray light, full of strange forms and structures.

Many phenomena repeat themselves. Here is a star like the Sun; there's another one, and another, and another. But there are always new wonders, too. We can find something extraordinary almost anywhere we look. Here is a pair of stars orbiting so close together that the gravitational pull of one on the other is significantly greater on their facing sides than their outward sides. Raw stellar material, scorching plasma, is streaming across the gap between them in a tug-of-war that may result in one star gobbling up much of its sister. In a remote corner of another galaxy is a giant old star just minutes away from yielding itself up to gravity's persistent embrace. When it does, its core, now transmuted to iron and nickel, will implode and collapse inward, only to crash into itself with a backlash that will blow the star apart. A great and brilliant supernova will ignite, perhaps to be glimpsed by some life-form on a planet orbiting another star in another galaxy.

And here is a black hole, streaking along a remarkable trajectory at hundreds of miles a second. Some earlier event, perhaps during its formation, or perhaps in some frighteningly close encounter with another massive object, has ejected it from the galaxy it once called home. Its long journey is taking it out into intergalactic space, a lonely, dark voyager. Over there is a double pulsar. Twin neutron stars, each the mass of two Suns yet less than eight miles across, race about each other like the ends of a dumbbell spinning in space. Both are spinning furiously, taking mere fractions of a second to complete one rotation. Each is beaming out intense radio waves. And off in a majestic spiral galaxy is a small rocky planet orbiting a moderately bright star. Its three small moons glide around, tugging at the great equatorial ocean on its surface. In the gently lapping waters on its distant shores, a thick green carpet is slowly growing, home to countless microscopic forms that scuttle about their business.

Our sack of universe stuff is huge, yet it represents a tiny, tiny fraction of the total observable universe. It also represents a very small slice of cosmic time strata. In the few hundred million years that it takes light to cross the sack, very little changes. Big hot stars have eaten through their hydrogen fuel and become bloated and old, or have exploded as gravity seeks to rearrange their equilibrium. New stars and planets have managed to form out of nebulous gas and dust. A few galaxies may have encountered each other, engaging in gravitational ballroom dancing that plays out over even longer timescales. Various small worlds have cycled through changes in environment, from ice ages to watery tropics. In the grand scheme of things though, on a truly cosmic scale, nothing much has happened.

› › ›

So here we have it: a small piece of a map of forever. If we could see the rest, it would span the largest scales of the observable universe, all 13.8 billion years of it, yet also be filled with the infinitesimal detail of every jagged crumb of rock, every porous cloud of gas.

In many respects, even our little map is still wildly incomplete.

Big gaps exist. We've been able to probe some locations, but not others. We've been able to probe only so far back in time. Nonetheless, it's a remarkable picture of the cosmos. It already takes in everything from the tenuous distribution of individual electrons, atoms, molecules, and microscopic dust all the way up through planets and stars. Then it proceeds onward to the great swarms that are galaxies, and farther to galaxy upon galaxy in clusters and superclusters, and finally to the walls, threads, and voids of the cosmic web. Some of it is delicately hued. The barest whispers of radiation emanate, whether in the ethereal realm of long-wave radio waves or the sparse but potent punch of the highest-energy cosmic particles—exotic subatomic crumbs whizzing through the cosmos in relativistic stasis, time apparently held almost still for them. Other parts are fiercely vivid, painted in the primary colors of intense infrared, ultraviolet, X-ray, and gamma-ray waves. Even the sea of photons from the early universe is extraordinarily rich: hundreds of them are thronging through every cubic centimeter of the cosmos.

It's a map of endless signposts and byways. What may appear to be dull, uninspiring little galaxies contain a myriad of treasures: new stars and planets, tangled magnetic fields, and intricate and always unique dynamics. Trillions of objects dance among themselves as gravity pulls the strings.

A map like this serves many purposes. It helps satisfy our deepest longings for a sense of order and place. It allows us to identify, name, and study specific objects and phenomena. It allows us to quantify their relationships to the cosmographic surroundings, and to compare and contrast them to other cousins. Galaxies reveal more when we can compare them to other galaxies, nearby and distant, with common or unusual characteristics. Stars of different compositions all follow a pattern described by fundamental physics, yet exhibit enormous variety in their day-to-day behavior. Our map offers a vision of the universe complete enough for us to try to apply our theories and models of the cosmos, testing them against each other to deduce the universal nature of matter and energy and the underlying behavior of spacetime itself.

As the map of our known universe has been put together with increasing speed over the past hundred years or so, helped along by work like that of Harlow Shapley in 1918, and by Edwin Hubble's later revelation that many nebule were actually distant galaxies, zooming away from us, we have been able to make some very bold statements. One that is critically important is the simple estimation of the population statistics of objects. Just as, on Earth, knowing how many humans there are, or how many trees or volcanoes, helps us assemble a description of our planet, understanding the nature of the cosmos hinges on statistics. With careful extrapolation, we can now estimate that the total number of galaxies in the observable universe probably exceeds 100 billion, and may be closer to 200 billion distinct systems. A single large galaxy such as our own Milky Way may contain upwards of 200 billion normal stars. Both for stars and for galaxies, the most numerous examples are also the smallest. About 75 percent of all stars in the Milky Way are less than half as massive as our Sun, and out in the universe, the majority of galaxies are classed as dwarfs, containing just a few hundred million individual stars.

The numbers are somewhat flexible; estimation is as much art as science in this case. Nonetheless it's a safe bet that there are 1,000,000,000,000,000,000,000 (10^{21} or a billion trillion) individual normal stars in the entire observable universe—and possibly ten to a hundred times more than that. It is a mind-bogglingly large number. It's interesting to note that the total number of human beings ever born (or at least counting back to 50,000 or 100,000 years ago) is often estimated as about 110 billion. So, roughly speaking, there are about 10 billion stars in the universe for every human being who ever existed.

We're not going to run out of stars anytime soon. But the really intriguing question is, why is it this number at all? Why do we look up at a night sky with this particular balance between light and dark, between the scattered points of brilliance and the blackness of space? The propensity for our universe to make stars, and how efficiently it makes them at any point in cosmic history, is what determines the number of stars we can see—whether within our own galaxy or in the galaxies beyond, and even in the sparse terrain of intergalactic space. Since we are the product of generations of stars forging heavy elements, and are reliant on the energy of a star—our Sun—for maintaining the surface environment of Earth, it is critically important to know what the recipe is for making the universe this way.

We've posited a deep connection between this question and the extraordinary phenomena that are black holes—a connection that implies that they play an active role in sculpting the universe. Our map of forever gives us a jumping-off point for following this line of reasoning. This is the universe as we see it today, yesterday, and through 14 billion years of history. The connections between this atlas and the gravity machines that have helped craft it are there for the taking. We just need to find them.

3

ONE HUNDRED BILLION WAYS TO THE BOTTOM

On a sweltering September day in 1935, a twenty-car motorcade ploughed along a dusty road, heading toward a desert canyon. President Franklin Delano Roosevelt was on his way to dedicate the Hoover Dam. Waiting for him were the two Winged Figures of the Republic, sitting solemn-faced on their black plinths on the dam's western side. Their muscular arms stretched up and fused with great blade-like wings of bronze, soaring thirty feet toward the crystalline blue heavens. To their south and east the world dropped away, a seven-hundred-foot-deep void opening up in the dry, rocky landscape. To the northeast, a glistening body of water lay between ancient cliffs. After arriving and being helped out of his car, Roosevelt took a good look around. For once he was genuinely speechless. The colossal, elegant structure surrounding him was staggering and overwhelming. He eventually began the official opening with a witty but heartfelt expression of appreciation for its monumental form: "This morning I came, I saw, and I was conquered . . ."

The Hoover Dam is a temple to gravity, its apex straddling the rocky border between Arizona and Nevada. The enormous forms of concrete and steel are poised in space and time, feeding off the energy of the matter in the Colorado River as it squeezes through this gorge. The dam is much more than a great engineering feat; it feels profoundly connected to the hidden glue and threads binding the universe together. It's perhaps little wonder that when the United States Bureau of Reclamation sought the opinion of the famous architect Gordon B. Kaufmann on the dam's design, he felt inspired to commission the organic curves of modernism and Art Deco that now characterize the structure, including the great winged sentinels. Set into the floor beneath these figures is an extraordinary map made of carved and polished stone. This great terrazzo mosaic of the stars and planets is constructed with exquisite precision, reflecting the celestial hemisphere as you would have seen it on the exact date in September 1935 when a humbled Roosevelt made his dedication. It's designed to serve as a clock, a navigation aid across the ages. If in a thousand years archaeologists come across this map, by applying their astronomical knowledge they will be able to date precisely when it was positioned in the ground. Language is not a barrier. The stars provide their own translation. It's impossible not to marvel at the wonderful sense of pride and optimism of the people who conceived and built the dam. Even postmodern cynicism is tempered in the presence of such a monument to human innovation and its cosmic connections.

The Hoover Dam is built around the very same basic principles that govern the behavior of matter anywhere in the universe. Standing on its great walkway, you are far closer to gravitational physics than you might suspect. The weight of water trapped against its wall by gravity represents an enormous store of energy. Trillions of gallons fill the vast reservoir of Lake Mead to its north. This mass of water not only provides the impetus that is used to drive the huge electrical generators at the dam's base, but also pushes against the convex face of the dam, transferring forces into the natural rock walls to either side and ensuring that the concrete seals tightly and securely. Following universal rules, the waters of the Colorado seek the shortest path to fall along, deep inside the Earth's distorted space and time. Here, in the cusp of land between Arizona and Nevada, is a beautiful and visceral example of a phenomenon that is critically important for understanding the cosmic nature of black holes.

This great dam can extract as much as 2,000 megawatts of electrical power from the water rushing through its base and spinning its turbines. This adds up to an astonishing 4 terawatt-hours roughly every year—enough to power a small city. By sealing off the natural gravitationally driven flow of water, it has created a literal mountain of liquid, poised to release energy as it flows downward. We call this power source hydroelectricity, and across our planet it has become a critical resource in helping us maintain our lives. In some places it is truly indispensable. Thousands of miles away from Nevada, the country of Norway is blessed with an extraordinary and beautiful topography of mountains, containing more than thirty thousand elevated lakes that tap directly into natural planetary cycles. Water vapor is lofted into the atmosphere through solar-driven evaporation, returning as rain to these high lakes and rivers, where gravity then accelerates it downward. Hydroelectricity in Norway generates a total of over 130 terawatt-hours per year, providing effectively all the country's electrical power needs.

We've learned how to extract gravity's impressive energy in these special circumstances, but to understand how this same principle works in the most extreme environments of the universe, we need to stretch our minds a little further. Gravity as we experience it here on Earth is a weak version of the phenomenon. Cosmic equivalents of the Hoover Dam, or Norway's mountain lakes, involve a whole different order of physics that builds on these terrestrial examples. I've already introduced the idea that gravity is really a side effect of the curvature of space and time, and now is the moment to tackle that concept head-on.

› › ›

When Einstein produced his work on special relativity in 1905, he profoundly altered the way in which we conceived of the universe.

Space and time, up until this point, had been considered separate entities by physicists, but Einstein revealed that they were intimately connected. He was able to reconcile our model of nature with what we saw around us by accepting the finite and unchanging speed of light and the invariance of the laws of physics in each and every frame of reference. But the tricky conceptual price was that space and time had to be both variable and inseparable. Together, they were something new that became known as spacetime. This implies a host of physical effects that hinge on the innate limitations to measurement and interaction implied by the finite speed of light. An object moving past us at a high constant speed appears shorter along its direction of motion. It also appears to have greater inertia. If it carries a clock, we see time on that object passing more slowly. Events that appear simultaneous to one observer might not to another who is moving past. In the wake of such a profoundly new worldview, it was little wonder that Einstein worried endlessly about tidying up the loose ends, especially those that would allow relativity to apply to *any* natural situation.

He was particularly concerned with gravity. According to Newton, the force of gravity accelerated objects. But that force depended on the distance between objects, and in Einstein's new relativistic spacetime, distance was a flexible quantity. An imaginary astronaut speeding toward a planet would measure a shorter distance between the space capsule and the looming world than would an observer standing off to the side. Each would then deduce a different gravitational pull on the astronaut—but this could not be true, since it would violate relativity's assertion that physical laws remain the same everywhere. Einstein was deeply puzzled. How could nature choose the right distance or frame of reference to make this all work? He knew that the old picture of gravity was missing something. Intuition told him that relativity must somehow apply in this case, too—that no one observer could be "special."

His first breakthrough, coming to him as he sat at work in 1907, was to realize that someone falling freely in a gravitational field would sense no acceleration. Carnival rides exploit this all the time. If you "free-fall" at an amusement park ride, your stomach feels like it's in your throat. It might be! For a brief moment you actually have no weight. Einstein's "thought experiment" suggested that a person's experience of free fall might be entirely equivalent to floating out in distant space, away from any gravitational pull. So, he reasoned, small frames of reference in a gravity field could indeed satisfy relativity's requirement that no single frame of reference is unique. For my example, in that instant of nausea at the carnival, you might as well be afloat in deep space. Any unfortunate experiments going on in your stomach are indistinguishable from those performed in a place free of gravity.

There was a problem, though. And it was a question of size. When objects fall toward the Earth, they all move in a straight line directed to the very center of the planet. This means that two opposite ends of an object are actually being pulled toward each other as the object descends. Imagine a gigantic blue whale falling toward the Earth—not because we want to damage it, but because through no fault of its own it's rather large and convenient for an experiment. Owing to whale aerodynamics, our blue whale falls horizontally. Now, both its head and its tail are falling along lines that point directly to the center of the Earth. But that means that as it descends, its head and tail are being pushed together, since the closer we get to the planet the smaller the distance is between these two radial lines. There's a name for this in classical Newtonian physics: the whale is feeling a gravitational tide. It also feels a stretching force between its belly and its back. The whale's belly is closer to the Earth, and so it feels a slightly stronger gravitational pull than the whale's back. So the poor whale is being squashed lengthwise and pulled apart bottom to top by tides. Whether it's more squashed or more stretched depends on the precise size of the planet and the size and shape of the whale.

This represents Einstein's dilemma. The effects felt by the whale are real, and are described easily by Newton's theory of gravity. Yet he knew that this same theory was not compatible with relativity. Einstein could make gravity and relativity work for tiny frames of reference where tides were too small to worry about, but a universal law must apply to any situation. The only way to reconcile the problem was to throw out Newton's theory of gravity and start over.

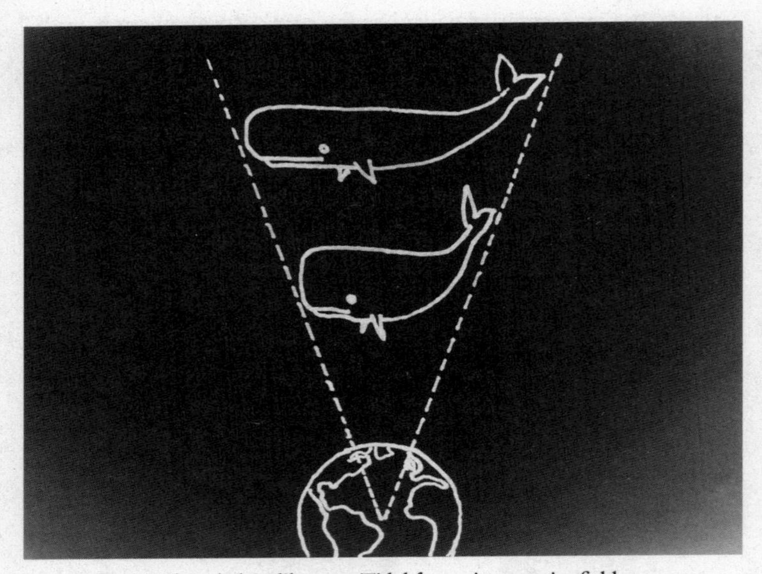

Figure 6. The whale's dilemma. Tidal forces in a gravity field squeeze the whale head to tail and stretch it top to bottom, as is well explained by Newton's theory of gravity. But that theory is not compatible with relativity, in which measured distances depend on the frame of reference. Einstein's solution was to rework the whole meaning of gravity. In the theory of general relativity, tidal effects are a direct consequence of the distortion of spacetime itself.

This was and still is a conceptual leap of such insight and colossal arrogance that it boggles the mind. It profoundly altered more than three hundred years' worth of fundamental physical research. So insistent was Einstein that relativity had to be correct that he ended up saying that Newton's theory of gravity was fiction. In its place, he posited that spacetime was itself flexible. A mass like the Earth curves spacetime around itself and toward itself. Each part of the whale follows the shortest path through this spacetime. The tidal effect it feels pushing its head and tail together is *entirely* equivalent to saying that spacetime is getting more bunched up toward the Earth. Similarly, the pulling between belly and back is because spacetime is stretching in the radial direction. To solve his problem, Einstein decided that gravitational tides and the curvature of spacetime are just different descriptions of the same thing. What everyone had been calling gravity is just the way objects move in this distorted spacetime. But, although he had found the conceptual solution to incorporating gravity into relativity, he still had to formulate a mathematical framework to describe what was now a *general* theory of relativity. This framework had to relate the curvature or distortion of spacetime directly to the mass doing the distorting.

This was an immense challenge, and it taxed Einstein to the limit. In late 1915, after a number of half-successes, he finally arrived at a mathematical description that completely encapsulated this new physics. He would not have gotten there without the work of many others, including those who essentially arrived at the very same point at the same moment. However, Einstein was the one individual whose incredible intuition and persistence broke through the layers that had obscured this improved description of the universe.

All those years of work can be summarized in what is referred to as the Einstein field equation, a form that is surprisingly but deceptively simple:

$$\mathcal{G} = \frac{8\pi G}{c^4} \mathcal{T}$$

Equations can induce panic in many of us, but there are features of Einstein's field equation that we can readily understand without going very deeply into its full complexity. It's just describing all the things we have already talked about. On the left-hand side, the symbol $\mathcal{G}$ is the description of the curvature of space and time. On the right-hand side, the symbol $\mathcal{T}$ contains the description of mass and energy in the bit of spacetime in question. So, for example, $\mathcal{T}$ might be thought of as the instructions that describe where a spherical mass is, how massive it is, and whether it's moving or rotating. $\mathcal{G}$ then contains the instructions for how you or I, or a large whale, will move in the distorted or curved spacetime around that mass. Those instructions naturally also contain the information about how coordinates and distances will work, and how time may be dilated. $\mathcal{G}$ is often known in technical terms as the Einstein *tensor* and $\mathcal{T}$ is known as the stress-energy tensor, easily remembered as stress is what you get when you distort or curve something.

The second feature to pay attention to is that cluster of symbols before the $\mathcal{T}$. In there is the familiar symbol for pi (π), which has a numerical value of about 3.141. The symbol G is the gravitational constant. This is a constant of nature that was also part of Newton's original formulation of the laws of gravity. It simply describes the strength of gravity relative to all other forces, and it's a small number: about 6.67×10^{-11}, with the unit being meters cubed per kilogram, per second squared. On the bottom (the denominator) is c^4, which is the speed of light multiplied by itself four times. That, by any standard, is a big number. It's about 8×10^{33}, or 8 followed by 33 zeroes, since light speed is about 300,000,000 meters per second. What this all means is that the bunch of constants in Einstein's field equation end up as a tiny, tiny factor, and the equation becomes:

$$\mathcal{G} = 0.0000000000000000000000000000000000000000002 \times \mathcal{T}$$

This isn't just playing with the numbers—it's telling us something very important: while spacetime can curve and is a flexible medium, it is also *extremely* rigid and stiff. It takes a lot of T to get a little bit of G. In other words, you need a great deal of mass or energy in a very small region to get appreciable distortion of spacetime. A large building or a big mountain doesn't cause us to swerve in our cars or topple off our feet as we walk down the street. Indeed, it takes all 13 trillion trillion pounds of the planet Earth to keep us stuck to its surface. Even inside this curved space we can still do the high jump or throw a baseball into the air. Little wonder that the Schwarzschild radius we encountered in chapter 1 has to be so tiny in order to create the event horizon of a black hole. Matter must be packed to an incredible density to produce the enormous stress on spacetime, the T, necessary to overcome its inherent stiffness and to warp the local environment enough to trap photons of light.

The distortion, or curvature, of spacetime described by G affects the fundamental geometry of positions, distances, and motions. I'll refer you back to figures 3 and 4 for a reminder of how to visualize this. The bunching up of coordinates that our falling whale felt nose to tail, and the stretching from top to bottom, applies to the propagation of light as well. The electromagnetic waveforms of photons emitted just at the outside edge of the event horizon are stretched effectively flat as they climb away. In other words, they are drained of all energy, and cease to exist. Hence we can never see them escape.

The stiffness of spacetime is connected to another phenomenon that is noticeably manifest around extremely dense collections of matter. When Karl Schwarzschild produced his solution to Einstein's field equation for a single spherical mass and gave birth to the idea of an event horizon, he had to assume that the mass was not rotating. Yet if we look out into the universe with our eyes and our telescopes, this is not the way most objects behave. We're well aware that the Earth spins about its axis. But the Sun also spins. All the planets and moons spin. We can measure the spin of other stars. The messy material around baby stars is rotating in great dirty wheels of dust and gas. Spiral galaxies rotate. Matter in the vast nebulae slowly drifts and moves. The periodic, often circular, motion of solid or gaseous material about a central axis is a universal phenomenon. This leads to a question that may seem obvious now, but it wasn't always. If black holes originally form from the remnants of big old stars, what happens to their spin?

This is particularly intriguing for two reasons. The first is that we've said that if a mass is packed within its Schwarzschild radius then the external universe receives no further information about it. No light escapes; no information or event can be transmitted outward. We can't ever know what's happening inside this horizon. The second reason is that spin is a property that is very hard to get rid of once an object has acquired it. The universe likes to preserve, or conserve, rotation. More accurately, what it likes to conserve is angular momentum. This is the mathematical product of the distribution of mass of an object with its speed of rotation. Even the most jaded physicist will refer to the simple analogy of an ice-skater to explain the concept, and it's an analogy that works well. When ice-skaters pull their arms and legs tightly together, creating a beautiful rotation, they can speed up to a dizzying rate and wow the judges. What they've done is to shrink the distribution of their mass by moving it inside a smaller radius. The universe must compensate by increasing the spin, because angular momentum is conserved by nature.

Now imagine taking an object like the Sun and compressing it to within its event horizon (its Schwarzschild radius). In its present form the Sun spins once around its axis in a regal twenty-five days, with some variation since it is not a solid body like the ice-skater. It is about 1.4 million kilometers, or 870,000 miles, in diameter. If it were to shrink to within its event horizon of about 6 kilometers, or 3.7 miles, it would have to speed up to a rate of spin at which each complete revolution would take about 0.0001 second—a ten-thousandth of a second. This sounds ridiculous, but we know that this is exactly the kind of thing that happens in nature. Neutron stars (the ultradense giant atomic nuclei left over from massive stars) can zip around in a few thousandths of a second, and as we've seen, these are just a short step away from being black holes. Based on what we know about real astrophysical objects, it seems inevitable that some black holes must form with enormous spin. But doesn't the event horizon block such information from us? This was a further challenge for mathematicians and physicists trying to get a grip on this revolutionary physics.

In the decades following Einstein's formulation of general relativity, many scientists worked on finding new mathematical solutions to the field equation. One that eluded everyone's efforts was a solution that would incorporate the spin of a spherical mass—the G that results from a spinning T. It was a tough nut to crack. Then, unexpectedly, in 1963, a young mathematician from New Zealand named Roy Kerr gave a brief lecture at an astrophysics conference in Dallas that changed all this. Kerr had done it. He had found a solution that went beyond that of Karl Schwarzschild, and included the possibility of a spinning object. Those who attended the meeting recall that most of the audience didn't realize they were witnessing a moment of pure breakthrough. People dozed, and some even got up and left. But those who stayed alert were awestruck. Chandrasekhar would later write of the implications of this mathematical discovery as "the most shattering experience" in his forty-five years of doing science.

Kerr's solution sparked a flurry of work. It quickly became clear that black hole spin was not only one of the few properties that the event horizon did not hide away, but that it would be manifest in a most remarkable fashion.

In essence, a spinning massive object produces the same kind of effect as a tornado. The stiff and rigid spacetime surrounding the mass gets dragged around. Just as in the ferocious winds of a twister, this means that anything in the region will get dragged around, too. There is nothing you can do about it. Even light approaching a spinning black hole will end up being dragged around and around instead of immediately traveling straight down. What is most remarkable is that because this happens just *outside* the event horizon, the property of black hole spin is visible to the rest of the universe.

This extraordinary characteristic led to further revelations. In 1969, the English physicist Roger Penrose argued that the energy held in a spinning black hole could be extracted. The essence of his idea can be seen by a simple example. Imagine you throw a large and crumbly brick toward the side of a black hole that is spinning away from you. At the moment the fragile brick enters the strongly dragging spacetime outside the hole, it splits into two pieces. One chunk moves on a trajectory head-on into the moving spacetime that causes it to drop to the event horizon and vanish. The other piece, however, moves in alignment with the whirling spacetime and manages to escape—like a surfer catching a wave. In Penrose's process, the escaping chunk of brick can be moving fast enough that it carries off more energy than the entire original brick had. That extra energy comes out of the black hole's spin.

Again, it is because the spacetime *outside* the event horizon is being dragged around that the cosmos can get its sticky hands on that energy. Matter and radiation can pass this energy off into the universe by the boatload. In theory, the maximum rotational energy that can be siphoned off from a fast-spinning hole is equivalent to about 28 percent of its mass converted to pure energy. This is almost fifty times more efficient than the Sun's production of energy by nuclear fusion in its core. In fact, black holes may be the ultimate in power-generating flywheels, and this possibility raises a critical question. Can the behavior of matter within the extreme curvature of spacetime around a black hole produce a smoking gun that actually reveals these environments to us?

›››

On Earth, we have learned to extract energy from matter falling in curved spacetime. The Hoover Dam is a wonderful example of this, as is any hydroelectric plant. In the case of the Hoover Dam, billions of gallons of water accelerated to high speeds push against the blades of huge turbines that convert the energy of motion into electrical current. Out in the universe, if matter falls into the curved spacetime around a mass—sometimes described as a gravity well—it too gains speed, and gains what we call *kinetic energy*. This accelerating matter can then collide or interact with other falling matter along the way. Like water pouring down a slide, it churns and froths as it splashes and crashes into itself. Some of the kinetic energy gets converted into other forms. Everything from photons to subatomic particles can be spewed forth by fast-moving matter with high kinetic energy. Not surprisingly, the amount of kinetic energy gained by falling matter increases with the amount of mass in a system. The amount of energy also critically depends on how far an object can fall, how close it can get to the bottom of a gravity well. This is a factor that, as we will see, places black holes apart from anything else in the cosmos.

Suppose we could mischievously drop a potted plant, perhaps a nice geranium, from the location of the Earth and have it fall toward the Sun, starting from a standstill and ignoring Earth's pull. The Sun is a long way off. It is as if we're dropping something down a 93-million-mile-deep pit. By the time our dropped pot reaches the outer solar atmosphere, it will have gained a considerable amount of kinetic energy. It will hit the visible surface of the Sun with a terminal velocity of about 370 miles a second. If the plant in its pot has a mass of one kilogram, or a little over two pounds, the kinetic energy it carries is even more astonishing. It is equivalent to the energy of 100 billion apples being dropped from one meter above the surface of the Earth. That's also equivalent to about twenty tons of explosive energy, enough to level a small town.

Yet this is proverbial peanuts in cosmic terms. Let's suppose that instead of dropping our pot into the Sun, we dropped it from the same distance (one astronomical unit, or AU) toward a white dwarf, the dense remains of a once-proud star. A moderately large white dwarf with the same mass as the Sun will have a radius that is about a hundredth that of the Sun. This is key to understanding the energy gained by falling into gravity wells. The deeper you can get, the dramatically greater the kinetic energy you will gain. In this case our little pot will hit the white dwarf with a hundred times more energy than it hit the Sun with, even though the total masses of the Sun and the white dwarf are identical. It will be moving at about 6,000 kilometers (3,700 miles) a second, or 2 percent of the speed of light, and will crash down with the energy of a 2-kiloton nuclear bomb. This is all from just being dropped and allowed to fall along the shortest path in the curved spacetime around the white dwarf. It's a cosmic game of water balloons, or flowerpots, dropped onto an unsuspecting passerby.

If we performed the same experiment on a neutron star, which is only about six miles in radius, the result would be even more extreme. With end velocities approaching 30 percent of the speed of light, we would have to modify our estimates of the final energy of impact to properly account for the effects of special relativity, according to which our nice little pot would appear to gain inertial mass as it sped up.

What about the ultimate extreme, a solar-mass object squeezed to a radius within its event horizon of three kilometers (1.86 miles), making a black hole? We may be giddy with anticipation, but we actually already know what will happen to our two pounds of pot, soil, and plant: falling from a distance, it will accelerate ever closer to the speed of light itself, reaching that ultimate velocity right at

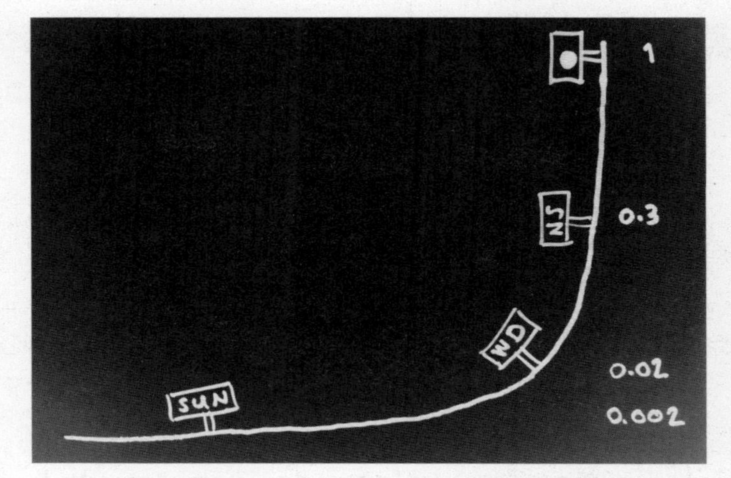

Figure 7. An illustration of the terminal velocity of objects dropped toward equal-mass astrophysical bodies from a great distance. The more compact the body, the deeper into its gravity well objects fall before hitting its surface, and the more they accelerate before they hit. The Sun is almost 435,000 miles in radius, and objects will hit it at 0.2 percent of the speed of light. A white dwarf of the same mass as the Sun is about 4,350 miles in radius, and objects hit it at 2 percent of the speed of light. A neutron star of this mass is only 6 miles in radius, and objects hit at about 30 percent the speed of light. The event horizon of a black hole of the same mass is less than 2 miles in radius, and the terminal velocity would technically be the speed of light itself (shown here as the value 1).

the event horizon. But there is no true surface for it to hit, no place to release all that kinetic energy. Furthermore, the distortion of space and time in the immediate vicinity of the event horizon becomes so extreme that what we might see as a distant observer becomes rather confusing. Our information arrives in the form of photons from the flowerpot that have climbed out of the fiercely curved spacetime around the hole. These are increasingly redshifted and diminished in energy as the pot descends to the horizon. Not only that, but the passage of time for the flowerpot appears ever slower to us as it approaches the black hole. That final explosive "pop" will never come.

Nonetheless, the mass will be moving at a tremendous speed on its way down, well before the event horizon and the shroud of these relativistic effects. If the black hole is spinning and sweeping the pot around and around, the result is further amplified. Should the pot intersect and collide with anything on the way, the potential exists for an enormous release of kinetic energy, converted into the motion of atomic and subatomic particles and electromagnetic radiation. Produced well before reaching the event horizon, these particles and photons can escape, surging back out into the universe. A crude analogy is to liken this to water draining noisily from a bathtub. As the liquid falls down into the drainpipe, some of its swirling kinetic energy is converted into sound waves, water bashing against molecules of air. The sound waves move faster than the water, and they escape. That gurgling sound we hear comes from the energy of the moving water converted to the movement of molecules in air. This movement is transmitted from molecule to molecule, like a line of falling dominoes, and the pressure beats against our eardrums. Finally, our ears convert those forces of movement into electrical impulses that flow off into our brain.

This generation of outgoing energy as matter moves in distorted spacetime is a defining characteristic of our universe. Mass stresses and curves spacetime around itself, and like the water falling through the channels of the Hoover Dam, matter falling into such a place can gain and release energy, lots and lots of it. It's a very efficient process, and it gets more efficient the more distorted spacetime is. Black holes represent an ultimate extreme of this, so compact that they pull the universe in after themselves and even drag it around and around as they spin. The next key part of our story is exactly *how* we detect the energy produced as matter approaches the gravitational drainpipe of a black hole. Without some equivalent of the noisy slurps of escaping bathwater, black holes would remain hidden away, lurking in the dark corners of the cosmos. Luckily for us, the real situation is very different.

›››

It is almost midnight in New Mexico on the eighteenth of June in 1962. From a viewpoint at the altitude of low Earth orbit, 140 miles above the White Sands Missile Range near Las Cruces and the Rio Grande, the dark mass of our planet fills the sky. Here it is utterly silent in the vacuum of space, and the last appreciable wisps of atmosphere are more than sixty miles beneath us. Off to the southeast an almost full moon hangs brilliantly in the blackness of space. The tiny points of stars seem to litter the cosmos. All is completely still. Then an object appears below our vantage point, shining brightly in the lunar glare. It gradually moves closer. A slender white cylinder, a rocket almost thirty feet long, is climbing up through the Earth's gravitational well. Its fuel is spent and it's now coasting nearly vertically, slowing down as it rises. Very soon it will reach the apex of its flight right where we are perched, and will then begin to fall back to the Earth. Apart from a few yellowish lights scattered across the planetary surface below us, there is nothing to give away the bustling civilization living there. Up here is space, and this small vehicle is only making a brief visit to sound out the depths.

Packed into the side of its nose are three Geiger counters designed to measure radiation, particularly the presence of X-ray photons. This skinny arrow of a rocket is spinning around twice a second, and its crude detectors sweep across the face of the Moon, sniffing for radiation produced as the Sun's light beats down on the lunar surface. There is not much. Then something happens. Away from the Moon, some 30 degrees of arc off in the seeming blackness of the sky, the counters begin to click faster and faster. Two, three, four times faster than they had before. X-ray photons are pouring out of a mysterious new place in the cosmos. The rocket catches this scent for only a brief few moments. Already it is beginning to slip back to

Earth, starting a tumbling descent that will drop it onto the desert landscape of New Mexico. It is enough, though. For the first time, humans have seen evidence of a place in the universe that is aglow with the fire of something fierce and alien.

This arrow-like rocket lofted high above New Mexico in 1962 was one of a pioneering array of experiments to seek out forms of light from the cosmos usually blocked by the thick atmosphere of Earth. It was a part of a radically new type of astronomy, undertaken from space. One of the participants in this fledging scientific effort was a young Italian named Riccardo Giacconi. Born in Genoa in 1931, Giacconi had already led a life colored by great changes in the world. Through the turbulence of Fascist-led Italy during World War II and then through his own personal struggles in the face of conventional and stultifying science teaching, Giacconi nonetheless grew into a highly skilled scientist. By the late 1950s he had moved to the United States, where he helped develop and build a variety of experiments to detect exotic and fast-moving subatomic particles. Then, as the space age began in earnest, he found himself immersed in the design and launch of experiments on small rockets. These were not much more than modified missiles. Known as sounding rockets, they could not attain orbit around the Earth, but they could climb a few hundred miles before falling back down. During their five minutes or so in space it was possible to carry out all manner of experiments, from detecting radiation to measuring magnetic fields.

That is why, on a summer's night in 1962, a set of radiation counters flew high above New Mexico. They were designed to probe the interaction of solar radiation with the Moon. As the intense stream of stellar particles hit the lunar surface, it was expected to scatter out X-rays and make the Moon the second-brightest source of this radiation in the sky compared to the Sun. The actual outcome was surprising and extraordinary. The brightest flood of X-ray photons that the rocket saw in space was not from the Moon at all. It came from an entirely different direction, toward the constellation of Scorpius, the Scorpion. What Giacconi and his colleagues had discovered was not generated in our solar system; it came from far beyond. This was the first sign of an entirely new universe, one unseen by human eyes. It was a regime of the most physically violent and energy-rich phenomena in the cosmos, the realm of what would come to be known as high-energy astrophysics. The mysterious new source of intense X-rays was given the name Scorpius X-1 and quickly prompted a flurry of effort to identify its origins.

A speculative picture emerged that astronomers have since confirmed. At the heart of Scorpius X-1, some nine thousand light-years distant from Earth, is an incredible system of two objects. One is a neutron star, a fearsomely dense core of nuclear matter. Its companion is a normal star, but that star is being eaten alive. The spacetime distortion around the neutron star siphons off stellar material from its companion, and as it plummets inward, kinetic energy is converted to other forms, including intense X-rays. It was obvious that if X-rays could be seen from this one corner of the universe, there must be myriad other places to look. In the years following the first detection of Scorpius X-1, increasingly sophisticated instruments were launched on sounding rockets in order to sniff for more sources of extraterrestrial X-rays. During one of these flights, in 1964, another new object showed up. This time it was an intense flood of X-rays from the direction of the constellation Cygnus, the Swan. Cygnus X-1, about six thousand light-years away from the Earth, was simply added to the growing atlas of the X-ray universe. But in 1970 a new type of X-ray experiment, an orbiting platform of radiation detectors, took a fresh look at Cygnus X-1 and found something incredible.

This fully space-based orbiting observatory was named Uhuru, the Swahili word for "freedom," in honor of the location of its launch platform in the Indian Ocean, just offshore from Kenya. This close to Earth's equator, space launches can exploit the planet's spin as an aid in getting to orbit, gaining an extra kick of speed for free.

For the first time, Uhuru allowed astronomers to stare at X-ray sources in the sky to their hearts' content, for far longer than the five minutes of the sounding rockets. They discovered that Cygnus X-1 flickered. The intensity of the X-ray photons changed rapidly several times a second, even as fast as a millisecond—a thousandth of a second. The only way this could happen was if the source itself was small—less than sixty thousand miles across. Otherwise, the finite travel time of photons from the near and far sides of any structure would blur out these variations. It's like listening to music playing simultaneously from one speaker nearby and another speaker hundreds of feet away. This far apart, and the sound is out of sync and discordant. Put the speakers close enough together and it all comes into crisp harmony.

The clearly seen flickering was an indicator that Cygnus X-1 was a highly compact source of X-rays. At the same time, the amount of energy pouring out was terrific. X-ray photons take a lot of power to generate. The scientists looking at the new data knew that if this radiation was coming from ordinary matter, it was being heated to temperatures of millions of degrees. Maybe, they thought, it was matter falling onto yet another neutron star. At that point, however, new astronomical measurements of the spectrum of visible light in this system revealed that Cygnus X-1 consisted of *two* objects. These bodies swung about each other every six days, like children holding hands and waltzing furiously across a playground. One was a giant blue star. The other was an enigma, except that its mass was more than *ten times* that of the Sun. As we saw before, a neutron star cannot be that massive and support itself within the curved spacetime it generates. The only plausible answer was that this mysterious companion was a black hole. But with no surface for matter to crash into, how, precisely, was the black hole producing the energy that was pouring forth?

The groundwork for finding the answer had already been laid independently and simultaneously by two scientists living on opposite sides of the Iron Curtain during the Cold War between the Soviet Union in the east and the Western world. Yakov Zel'dovich was a brilliant physicist and a key architect of the USSR's nuclear weapons program. Edwin Salpeter was a brilliant astrophysicist who had been born in Austria, obtained his education in Australia and England, and finally settled at Cornell University in upstate New York. In 1964, Zel'dovich and Salpeter had both realized that there were specific ways in which matter could become ensnared by black holes that would result in incredible violence. The idea came from the known behavior of gas subjected to great speeds and collisions. As matter fell inward or was *accreted* by a black hole, it would crash and pile up against itself. In technical terms, it would be "shocked," much like the way a supersonic aircraft "shocks" the air around it to create a sonic boom. Temperatures in the gas would reach millions of degrees, and this would light up the black hole's vicinity with X-ray photons. Armed with these ideas, astronomers came to the conclusion that Cygnus X-1 really does contain a black hole. It was the very first clear example of matter gurgling as it plunges through warped spacetime toward a singularity.

The 1960s and '70s were remarkable decades for other branches of astrophysics as well. While Giacconi and his colleagues were busy launching their sounding rockets and satellites into space, scientists were also now exploring the cosmos at the opposite end of the electromagnetic spectrum. In this other realm, there were increasing signs of phenomena even more monstrous than Cygnus X-1 out in the distant universe. To appreciate this we need to follow a different astronomical story for a while, one that began somewhat earlier.

> > >

The field of radio astronomy was born serendipitously in the 1930s. While working for the Bell Telephone Laboratories in New Jersey, the physicist Karl Jansky noticed a curious signal in a new type of shortwave antenna he had built. Jansky had been set the task of trying to understand sources of radio emission that might interfere with Bell Labs' plans for a transatlantic shortwave radiotelephone service. His antenna was a large box-like construction of wood and stiff metal cable about a hundred feet long and twenty feet in cross section that sat on four Model T Ford car wheels. With some muscle power, the whole thing could be rotated on a circular track, and the antenna could be repositioned. By moving the antenna, Jansky could crudely locate where sources of radio noise were coming from.

The antenna picked up nearby thunderstorms with huge electrical arcs that sent out shortwave radio waves. It could pick up distant thunderstorms as well. And it also picked up a faint hiss of something else. Jansky carefully waited, monitoring this noisy static and watching it vary in intensity over the following weeks and months. Finally he was able to figure out where it originated. It wasn't coming from Earth at all. It came from a direction toward the constellation of Sagittarius, which was also the direction of the center of the galaxy. He realized that whatever was generating these radio waves was an utter mystery. Something was lurking out there among the stars.

When he published his results on "extraterrestrial electrical disturbances," it caused a bit of excitement—even *The New York Times* ran a story on it in May 1933. Jansky was eager to build a better antenna to find out what was going on, but Bell Labs wasn't having any of it. Having ascertained that this source of radio noise was something they would just have to live with, they sent poor Jansky back to work on other projects. No one knew what the origin of this radio emission was, but it acted as the seed for an entire field of discovery that would unfold over the following decades.

By the late 1950s, this field had developed to the point where astronomers had custom-built radio telescopes in the form of huge metal dishes and antennae arrays scanning and surveying the sky. They had found many sources of radio emissions in the universe, including the bright one toward the center of the Milky Way that

Jansky had managed to catch a whiff of. From electrically charged gas to planets, stars, and other galaxies, the cosmos hums with natural radio noise. Now scientists were hunting for new exotica, and hundreds of distant objects were being detected. In among those was a peculiar category of incredibly small but bright sources. Astronomers went to their visible light telescopes and took photographic images to try to figure out what these were. All they could see were tiny star-like blue dots where the radio waves were coming from. But splitting and dispersing the visible light out into a spectrum made things even more confusing: spikes and bands of light showed up at wavelengths that were hard to match up with the signatures of familiar phenomena. These mysterious objects looked like stars, but their fingerprints were all wrong.

Finally, in 1962, a series of careful astronomical measurements by a number of scientists in Australia pinpointed a particularly bright example of this mysterious class of object. The precise location soon enabled the Dutch-born astronomer Maarten Schmidt to target it with the giant two-hundred-inch-wide telescope on Mount Palomar in Southern California, successfully capturing a clean and precise spectrum of the object's visible light. Schmidt had immigrated to the United States a few years earlier to work at the California Institute of Technology and was already known for his studies of how stars formed from interstellar gas. As he pored over the photographs of dispersed light that had been taken of the unknown point of brightness, Schmidt began to realize what he was looking at. He knew that atoms of hydrogen emit and absorb light at very specific wavelengths. It's like a set of unique keys. The spectrum in front of him had these key marks, but they were all shifted downward toward redder colors, making them difficult to recognize. The simplest answer, in his view, was that this object was moving away from us at an incredible rate of speed: about thirty thousand miles per second.

Schmidt knew that according to special relativity the photons of a fast-receding source would be shifted to lower energies, a Doppler effect that would move the wavelengths of the spectral keys. That implied two things. First, to appear to be moving this fast, the object had to be caught up in the expansion of the universe and was literally about 2 billion light-years away. Second, if it was that far away, then for us to see it at all it had to be pumping out energy at an enormous rate. Schmidt ran the numbers by hand and arrived at an astonishing answer. This object was a trillion times as luminous as the Sun. It was pumping out as much energy as all the stars contained in a hundred normal galaxies. An object like this was beyond belief. It seemed impossible, and was so shocking that Schmidt recalls telling his wife that evening that "something terrible" had happened at work. The foundations of astrophysics had been well and truly shaken.

In the following years there was raging debate about what these "quasi-stellar radio objects" could possibly be. People eventually shortened that awkward handle to "quasars," but the mystery remained. By this time astronomers had found that many recognizable galaxies were also strong sources of radio waves. In these systems there was often no obvious single bright source; instead there were pairs of vast cloud-like "lobes" of radio glow. These were like nearly symmetrical dumbbells interleaved with the stars of the galaxies. They were invisible to normal telescopes of glass and brass, but strikingly obvious with radio dishes. Extending as far as a few hundred thousand light-years, through and beyond the galaxies containing them, these extraordinary glowing structures also required enormous amounts of energy to produce. When people first computed just how much, it seemed implausible for any known physical process. It was comparable to directly converting a million or more times the mass of the Sun into pure energy. The theory of relativity showed that energy and mass were equivalent through the famous expression $E = mc^2$. But there were seemingly few ways for nature to make this conversion with the efficiency required to explain what the radio astronomers were seeing. A source like nuclear fusion was woefully inadequate.

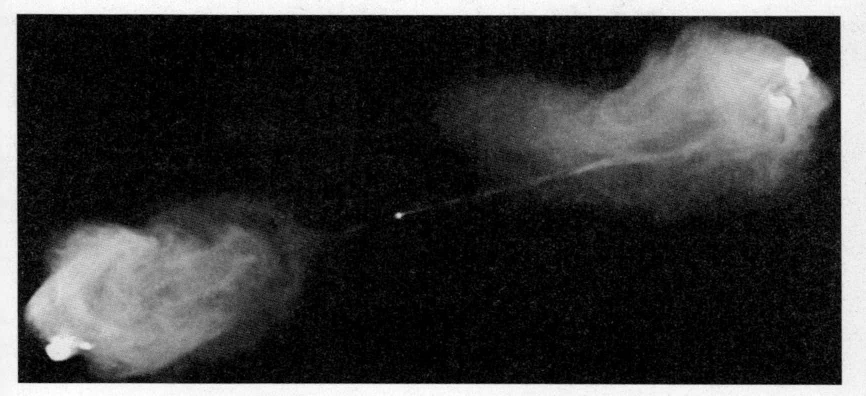

Figure 8. A modern image of the radio-wave light emitted from a galaxy that is 600 million light-years away. This remarkable object, known as Cygnus A, was discovered by radio astronomy in 1939. Early measurements simply showed a basic dumbbell structure extending across six hundred thousand light-years. With time, better and better images revealed the incredible thread-like structures between these huge clouds of what were eventually recognized as seething-hot electrons. The galaxy of stars is invisible in this image but lies between the clouds.

Even earlier in the 1930s, astronomers using optical telescopes had also recognized that many otherwise unremarkable galaxies showed evidence of extremely bright and hot spots toward their centers. In a few, there were even signs of curiously straight rays or "jets" of light emanating for thousands of light-years from these peculiar nodes. Puzzles abounded. The universe was once again full of strange and contrasting phenomena. Many demanded energy sources that far exceeded anything possible from chemical or even nuclear reactions. Could there possibly be a link between all these things?

The pieces were indeed all falling into place. Someone just needed to connect them. A number of incredibly compact and energetic environments seemed to exist in our own galaxy, with Cygnus X-1 emerging as the prototype. Mysterious radio emission was emanating from the center of the Milky Way. Radio astronomy had also found not just hundreds of distant sources across the sky, but also signs of the strangest forms and structures, great zones of radio emission spanning thousands upon thousands of light-years and containing colossal amounts of energy. Visible light revealed bright cores in many galaxies, together with some remarkable glowing ray-like prominences. It also seemed that these features might have a common point of origin with the mysterious and incredibly luminous quasars. But the power source for these phenomena was a huge puzzle. The generation of energy by matter falling toward extremely compact objects certainly met the efficiency requirements. But the kinds of examples starting to be discovered in the Milky Way were on a tiny scale compared to what was being seen in other galaxies and from the mysterious quasars. This couldn't be the answer, unless the black holes were enormous, millions to billions of times more massive than the Sun.

The original ideas of Zel'dovich and Salpeter would lead to the next conceptual leap, one that took this physics to a whole different scale. In 1969, a paper appeared in the scientific journal *Nature* that provided a single clean and elegant solution to the nature of the distant quasars and radio sources. It also laid out a new vision of the intimate relationship between black holes and galaxies. Its author was Donald Lynden-Bell, then in residence at the Royal Greenwich Observatory in Herstmonceux Castle in Sussex, some fifty miles south of London. Lynden-Bell was born in 1935 in Dover, southern England and had studied astronomy at Cambridge and worked at Caltech. Now he sat in the beautiful Tudor castle that some twenty years earlier had been purchased to house the Royal Observatory and its staff.

An incredibly productive and fluent worker with mathematics and physics, Lynden-Bell has made contributions that span much of astronomy. His influential paper from 1969 is full of provocative and exuberant language. Even the first sentence describes the great radio lobe structure in some galaxies as a "dead or dying quasar." He describes the Schwarzschild radius as the Schwarzschild "throat." The slightly greater distance beyond which no stable orbit can exist around a black hole becomes the Schwarzschild "mouth." But although his language was vivid and colorful, Lynden-Bell wasted no time getting to the point.

By the end of the first paragraph he had laid out exactly how the giant radio-emitting structures seen across the universe seem related to "dead" quasars, how colossal the energy output had been from these compact regions, how gravitational energy is more than competitive in this case with stellar nuclear fusion as the energy source, and how there must be dead quasars in our area of the universe. Behind all this: supermassive black holes ranging from 10 million to a billion times the mass of the Sun. It's a heady mix. By this time in 1969, there had already been discussion that the very distant quasars somehow represented a particular stage in the evolution of structure in the universe. Perhaps over the cosmic ages they morphed into the great radio-emitting lobes and clouds that were now being charted. Because we could only find the brighter and nearer quasars, there was a good chance we saw merely one in every thousand of the original objects. Lynden-Bell realized this implied that the true number of power sources for quasars in the universe was comparable to the number of galaxies. Matter accreting onto black holes a billion times the mass of the Sun could provide enough energy to be this power source. The logical conclusion was that the best place for these supermassive black holes was inside virtually every galaxy. At every galactic center, the great density of stars could provide ample fuel for the enormous appetites of black holes.

It's a work of great physical intuition, both outrageous and flawless. It goes on to present a mathematical description of how these giant black holes could eat matter and liberate energy. We'll encounter this physics again in more detail. Most extraordinary of all, though, is not how this idea was presented, but what its implications were.

If this hypothesis were true, then the universe would not only be full of small black holes that came directly from the deaths of large stars, but it would also be full of supermassive giant black holes. At the center of each galaxy—more than 100 billion of them, spread across almost 14 billion years of cosmic history—there must be black holes with event horizons that are tens of millions of miles across. This is a tall order to fill. While Lynden-Bell's elegant hypothesis was consistent with many facts, it wasn't immediately embraced by everyone. Arguments and disagreements swirled. Some astronomers felt that admitting giant black holes into our picture of the universe was too far-fetched. How was it possible for galaxies to contain these? How could you even grow such objects? Those dissenters suggested instead that vast agglomerations of stars dying as explosive supernovae could be responsible for the flood of energy from galactic centers. Others questioned whether the great velocities and apparent distances of quasars were real or an illusion of the gravitational redshifting of light. However, as time went by, more and more evidence mounted for precisely the kind of compact and incredibly dense regions at the hearts of galaxies across the universe that could only be matched by colossal black holes.

Like the gurgling of water rushing down a drain, energy was pouring out of these places just as matter was pouring in. Astronomers realized—just as Lynden-Bell had suspected—that this was not a sustainable situation. Objects like quasars would shut down due to starvation. But it also became clear that far from simply "dying" in obscurity, supermassive black holes could assume a variety of states. Sometimes they would still be eating, gently grazing and slowly growing in mass and releasing energy. At other times they would be quiet, sleeping off the previous few millions of years of food. Exactly how they ate could vary. Exactly how energy was released could vary. What we glimpsed of the process was also highly subjective, depending acutely on the chance arrangement of structures around these holes and the direction of our viewpoint.

To understand all this we need to take a very ambitious field trip. Our quest? To find out what really happens around a massive spinning black hole as our universe drains into it.

4

THE FEEDING HABITS OF NONILLION-POUND GORILLAS

Once upon a time there lived a great monster. It made its home deep inside a castle that was deep inside a huge forest. No one had ever seen the monster, but over the centuries and millennia there had been clear signs of it stirring. Legend told that it trapped all things that came near. In its lair even time itself became sticky and slow, and its hot blue breath would burn through the strongest shield. Few dared to venture into its realm. Those who did either returned empty-handed with wide-eyed tales too strange to believe, or never came back at all. If you stood on the highest mountains in the land you could peer across the treetops and just see the haziest of outlines of the monster's castle. Sometimes you might see a few strange clouds hovering over it, as if they were caught in a great swirl of atmosphere, and at night there might be an eerie glow reflected off the cool air. For years you've wondered about this enigmatic place and the monster within. Finally you decide that there is nothing else to be done but to go on your own quest, your own search for a glimpse of the beast. In this particular tale your starting point, and home, is our solar system, and the monster's castle is deep in the galactic heart.

At first the going is easy on your journey. The stars are familiar and friendly. Out here in the Orion spur of the great spiral disk of the

Milky Way, stellar systems are spaced with an average of about five to ten light-years between them. Finding a comfortable path through is not difficult. Even the rivers of dusty darkness between the galactic arms are easy to cross, and traveling the first twenty thousand or so light-years is a breeze. After a while, though, things begin to change. This is the beginning of the galactic axial hub. Like the distorted yolk of a huge fried egg, the central region of the galaxy inside about four thousand light-years is a gently bulbous but elongated structure. It contains a far higher density of old red and yellow stars than out in our suburbs. The woodlands begin to thicken up in here as we ease our way toward the inner sanctum. More and more stars begin to block the way, and we are constantly shifting our path in order to slide through.

Pressing on, we finally enter the true galactic core. Some six hundred light-years across, this interior forest is densely packed with stars buzzing around in their orbits. Compared to home, the skies are coated with star after star after star. At the edge of this core, where we first enter, stars are packed together a hundred times more densely than around our solar neighborhood. At the very middle, there are hundreds of thousands more than we are used to. The going is extremely tough and slow, and it gets worse and worse as we descend inwards. This is the oldest undergrowth, part of the ancient barrier to the center. Something else exists in here, too. A rather piecemeal and shabby disk of material encircles the entire core, made of hydrogen gas clouds. It blocks the view from some directions, and as we move farther down, another structure now begins to reveal itself. There is a flattened ring of gas rotating about the very center of the galaxy. It's composed of atoms and molecules, and it is unlike anything else in the Milky Way. It is a rich and substantial formation, a hundred times denser than a typical nebula. Its outer edge is still some twenty light-years out from the galactic center, but its inner lip descends to within only about six light-years. Tilted at a rakish angle to the plane of the entire galaxy, it spins at about sixty miles a second. Most of it is hydrogen gas, but nestling in among this pure stuff are other compounds: oxygen and hydrogen in simple combination, molecules of carbon monoxide, and even cyanide. Every hundred thousand years or so, the inner part of this molecular ring makes one complete circuit around the center of the galaxy. This impressive structure at first looks serene, but closer inspection reveals the scars of terrible violence. Some great cataclysm has recently blasted the ring, pushing some of the gas into clumps and lumps and scorching other parts. It is a strange and ominous gateway.

Moving cautiously inside the ring, we take stock of what is happening around us. We are within an incredibly dense and constantly moving swarm of stars. It seems like chaos, yet through this noisy buzz we can see something distinctly peculiar happening up ahead. We pause in flight to watch as several of these innermost stars move along their orbits. Remarkably, these orbits are not only *around* something unseen ahead of us at the center, but they are extraordinarily fast as the stars swing by that invisible focal point. One star whizzes through its closest approach at velocities approaching 7,500 miles a second. That's astonishing, considering that our homeworld, Earth, orbits the Sun at less than twenty miles a second, and even the planet Mercury moves at barely thirty miles a second. For the star to achieve an orbital velocity of that magnitude, it must be moving around a huge mass. We perform the calculation. Deep within a tiny volume at the galactic center is an unseen *something* that is *4 million* times more massive than the Sun. There is nothing else this dark body can be except a colossal black hole.

How we have come to build this detailed picture of the environment at the center of our galaxy is a tale of technological prowess and skilled insight. One of the greatest achievements of astronomy in the late twentieth century and early twenty-first century has been the discovery that our own galaxy, the Milky Way, harbors a supermassive black hole at its center. It provides a vital context for the rest of our story, and a key reference point. But there are still limits to how much detail we can see when we peer this deep into the inner galactic sanctum. At present we have to rely on a number of indirect astronomical phenomena to tell us more. For example, tenuous hot gas is being measurably expelled from this tiny region. X-ray photons are also streaming out, and roughly once a day they flare up and brighten by a hundredfold. It's tempting to imagine that somewhere inside this central core are moths flying too close to an open flame, and sometimes we see their unfortunate demise. Altogether these characteristics represent clear signs that matter is sporadically entering the maw of a brooding monster.

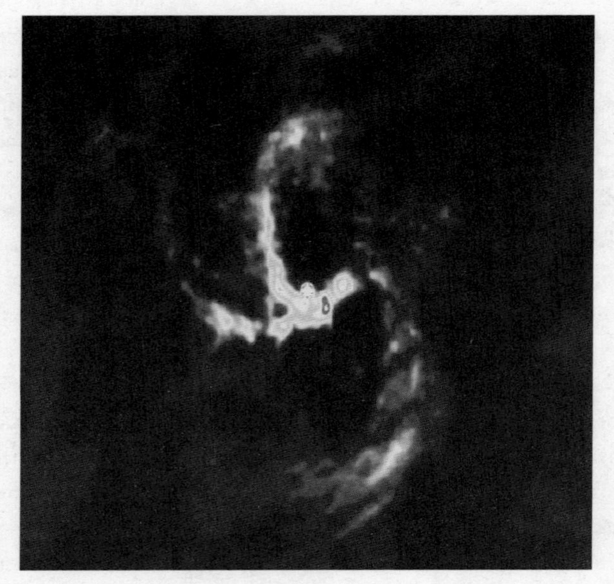

Figure 9. The innermost region of our own galaxy mapped at microwave frequencies. This image, spanning approximately twelve light-years, reveals an extraordinary structure of irradiated gas centered on a bright object that astronomers associate with the central massive black hole. As the image suggests, this gaseous structure is in motion around and toward a central point.

We see another signature in the great loops of magnetized gas that surround this whole region, aglow in radio waves that flood out into the galaxy. They are part of the very same extraterrestrial radio signal that Karl Jansky first saw in the 1930s with his simple radio telescope in a field in New Jersey. Yet despite all this activity, the black hole at the center of the Milky Way is operating on a slow simmer compared to the brilliant distant quasars that can shine as brightly as a hundred galaxies. It's a brooding, hulking beast, not a blazing pyre. But to really place it in context, we should size things up and compare this local environment to the rest of the cosmos.

To do that, let's return briefly to our map of forever, still contained in the sack that was delivered to the doorstep two chapters ago. In our neighborhood of the universe, encompassing a mere 6 billion years or so of light travel time, the intensely bright quasars occur in only about one out of every hundred thousand galaxies. In other words, they are extremely rare creatures. For that reason, we should not be too surprised that the Milky Way isn't one of the galaxies that contain a quasar. Those other galaxies with great radio lobes and ray-like jets extending outward are even more rare; the most prominent examples are over 10 million light-years from us. But at greater distances, further back in cosmic time, the situation is very different. In fact, between 2 billion and 4 billion years after the Big Bang, fiercely energetic quasars were a thousand times more common. We think that roughly one in a hundred galaxies held a quasar in its core at any moment. This was a golden age for these objects, powered by the voracious appetites of supermassive black holes.

No single quasar lasts for very long, however. With monumental effort, astronomers over the past several decades have surveyed and studied these enigmatic objects, and piece by piece they've reconstructed their history. Like paleontologists building the skeletons of long-gone creatures and covering them with reconstructed flesh, so too have astronomers rebuilt the lifestyle of the supermassive black holes that drive quasars. We find that a typical quasar will only light up for periods that last between 10 million and 100 million years, a tiny fraction of cosmic history. Because of this, we know that more than 10 percent of all galaxies in the universe have actually hosted a brilliant quasar during their lifetimes. It just means that wherever or whenever we look, we never get to see them all switched on at once.

But why do quasars die out with cosmic time? It is a question that remains unresolved. Even this basic description of the cosmic distribution of quasars is the result of decades of intense research. (The history of that effort is a fascinating one, but a story for another day.) We can, however, make some reasonable speculations about the life cycles of quasars. First, they are powered by supermassive black holes that, as they devour matter, produce an output of energy far greater than in other environments. The electromagnetic shrieks of material falling into a black hole are what we see during this process. This suggests that the enormous energy of quasars is deeply connected to the availability of consumable matter and the rate at which it is being consumed. The more matter falls in, the bigger the hole can become, and the bigger the hole, the more energy it can extract from that matter. Eventually, though, this material seems to run out. Quasars live fast and big and die after a blaze of glory that must depend acutely on the detailed nature of matter consumption by supermassive black holes.

The most distant quasars we know of (going back to within a billion years of the Big Bang) are typically also the most luminous. In other words, as the cosmic clock ticks, and new quasars come and go, they gradually become dimmer. The astronomical jargon used for this is "downsizing." (Who says scientists don't have a sense of humor?) All quasars, however, from the brightest to the faintest, are powered by the most massive of the supermassive black holes. They are the elite—the big guys. They also occur in the bigger galaxies in the universe. This is an important connection to make, because it begins to tie the evolution of supermassive black holes to the evolution of their host galaxies, their great domains.

Indeed, astronomers have found something else peculiar and critically important going on in galaxies. The mass of their huge black holes is generally fixed at one-thousandth of the mass of the central "bulge" of stars surrounding the galactic cores. These are typically the old stars that form a great buzzing cloud around galactic centers. Sometimes that central cloud can even dominate the whole galaxy. Careful astronomical measurements have revealed that a galaxy with a big bulge of central stars will also have a big central supermassive black hole, and a galaxy with a small bulge will have a smaller black hole—according to the 1,000:1 mass ratio. But while this relationship is strikingly clear in many galaxies, it is not entirely universal. For example, the Milky Way is pretty much "bulgeless." Its central stars are in more of an elongated block or bar, not a swarm thousands of light-years across. And, as we've seen, our own supermassive black hole is a comparatively petite monster of 4 million times the mass of the Sun. By contrast, the nearby spiral galaxy of Andromeda has a great big bulge of central stars and contains a supermassive black hole that we think is 100 million times the mass of the Sun, neatly fitting the expected size. Why there should be this relationship between central stars and black holes is a mystery at the forefront of current investigations. We will find it to be of the utmost importance as we dig deeper into the relationship between black holes and the universe around them. But the next step in following this story is to get our hands dirty again with the business of feeding black holes.

We can make a number of broad arguments to describe how energy is produced from the distorted spacetime surrounding dense concentrations of mass in the cosmos. I made some of those in the previous chapter, and emphasized the power involved. The idea certainly sounds feasible: there's plentiful energy to spare, but specific physical mechanisms are needed to convert the energy of moving matter into forms we can detect. Otherwise, it's like stating that burning gasoline releases a lot of energy and therefore an engine could be driven by gasoline. That might be true, but it doesn't demonstrate how an internal combustion engine works. In our case, the processes of energy generation and conversion are particularly complicated because of the exotic nature of black holes. Unlike an object such as a white dwarf or a neutron star, a black hole has no true surface. Matter that gets close to the event horizon will essentially vanish from sight for an external observer. There is no final impact onto a solid body, no final release of energy from that collision. So whatever is going on just outside the event horizon is absolutely critical to understand.

The early work on black hole energy generation by Zel'dovich and Salpeter in the 1960s, as well as that of Lynden-Bell, led to a number of theories about the mechanisms that could be at play. These involved a phenomenon known as accretion—the feeding of matter onto and into a body. But observation of the universe suggests that other things are going on as well. Something is responsible for producing the enormous energy-filled structures emitting radio waves from within galaxies, as well as the strange ray- or jet-like features emanating from galactic cores. In this case, the bizarre spinning ring of material that we find surrounding our own galactic center actually offers a general clue to one piece of the puzzle. In order to see why, it's time for us to properly consider the outrageous eating habits of black holes.

Although matter can fall straight down onto objects like planets, stars, white dwarfs, neutron stars, or black holes, in general it doesn't. What it does tend to do is enter into orbits. One way to think about this is to imagine a swarm of nearsighted bees flying across a field in search of a good nectar-rich flower. One such happens to be in the middle of their path, its bright petals giving a bee-friendly come-hither. A couple of lucky bees are lined up just right, and as the flower looms into their blurry vision, they simply land on it with a splat. The other bees, off to the sides, only barely notice something and have to swing their flight paths around to circle before coming in to land. In a sense, matter moving through curved space does the same kind of thing. If it's not perfectly on track to the very absolute center of mass of a large object, the most bunched-up point of spacetime, it will tend to loop around and orbit. As we've seen, all matter tries to follow the shortest path through spacetime, but if that underlying fabric is warped then so too will be the path. If the components of that incoming matter can also bump and jostle each other, they can further rearrange themselves. Atoms and molecules, even dust and bigger chunks of material, will settle into orbiting a massive body in a flattened, disk-shaped structure. We see this occurring everywhere in the cosmos. The arrangement of planets in our own solar system is an excellent example of this phenomenon. The flatness of their orbits reflects the disk of gas and muck that they formed out of some 4.6 billion years ago. The rings we see around Saturn are another example. Time and again, matter captured by the influence of a dense and massive body ends up swirling into an orbiting disk. It certainly seems that the same thing must happen around a black hole.

But if a black hole just swallows matter up, light and all, then how does it produce energy? The trick is that when matter forms a disk around the hole, the material in the disk rubs against itself as it swirls around. It's like spinning a stick against another piece of wood to start a fire. The pieces of wood are never perfectly smooth, and so friction between them results in the energy of the spinning motion being converted into thermal energy, and the wood gets hot. In an orbiting disk, the outer parts move much more slowly than the inner parts. This means that as the disk goes around and around and around, friction between the bands of moving material transfers the energy of motion into heating the matter. This has one very direct consequence: when you hold a hand on a spinning bicycle tire, the friction causes the tire to slow down and your hand to heat up. The same thing happens in the matter disk. The heated material loses orbital energy and spirals inward. Eventually, it gets to the event horizon and is accreted into the black hole, and it vanishes, sight unseen. But on the way toward that point, friction converts some of the tremendous energy of motion into photons and particles.

Figure 10. An artistic impression of a disk of material orbiting a black hole and glowing with light. In the background is a vista of stars and galaxies. To simplify things, the disk of matter is shown in a very pure state: no dust or other debris, just thin gas. It becomes denser and hotter as it swirls inward, heated by friction. At the very center is the dark event horizon, and the light in its near vicinity is bent by passing through this extremely distorted spacetime to form what looks like an eye. In fact, we're seeing the light of the disk that would otherwise be hidden from us on the far side of the hole, curved around as if by a giant lens.

Exactly what causes this friction is still a significant mystery. The force of atoms bumping randomly into one another simply doesn't suffice to explain what we observe happening out in the universe. Ripples and whirls of turbulence in gas may help roughen the frictional forces within the inner speedy parts of a disk, but they too are not quite enough. It may be that magnetic fields produced from the electrical charges and currents of material in the disk act like a great source of stickiness to produce the necessary friction.

Whatever the precise cause, there is absolutely no doubt about what happens when matter is ensnared this way. As it spirals inward through the disk, the friction generates huge amounts of thermal energy. Toward the inner regions, an accretion disk around a supermassive black hole can reach fearsome temperatures of hundreds of thousands of degrees. Powered by the huge reservoir of gravitational energy from the curved spacetime around a supermassive black hole, the matter in a single disk can pump out enough radiation to outshine a hundred normal galaxies. It's the ultimate case of friction burn. As Lynden-Bell originally saw in 1969, this is an excellent match to the energy output astrophysicists have seen in the brilliant quasars and inferred from the great structures of radio emission from many galaxies. This mechanism is also tremendously efficient. You might think that such a prodigious output would require a whole galaxy's worth of matter, but it doesn't. An accretion disk around a big black hole needs to process the equivalent of only a few times the mass of the Sun *a year* to keep up this kind of output. Of course, this adds up over cosmic time spans, but it's still a remarkably lean-burning machine. And there's even more going on, because spacetime around a black hole is not of the common garden variety.

We've touched on the effect a spinning mass has on its surroundings, the tendency to drag spacetime around like a twister. This phenomenon was one piece of the mathematical solution that Roy Kerr found to Einstein's field equation for a spinning spherical object. It's actually a more general description of mass affecting spacetime

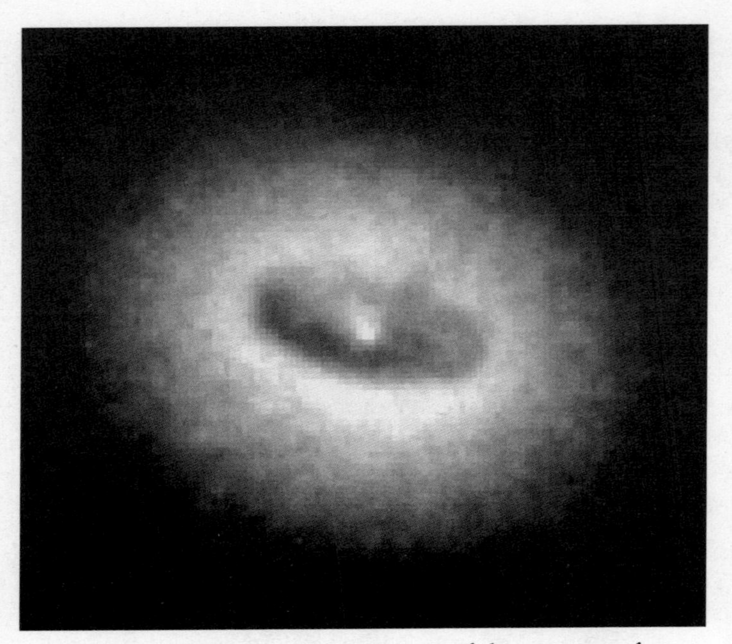

Figure 11. A Hubble Space Telescope image of the very center of an elliptical galaxy known as NGC 4261 that is 100 million light-years from us, still within our general cosmic "neighborhood." At the pixelated limits of even the Hubble instruments, this image shows a darker disk of thick gas and dust lying within the light of stars at this galaxy's core. The disk is tilted by about 30 degrees toward us and is some three hundred light-years across. It surrounds a supermassive black hole 400 million times the mass of our Sun (100 times the mass of the black hole at the center of the Milky Way). This material is slowly feeding into the bright disk of accretion-heated, rapidly orbiting matter seen as a point in the very center. That innermost disk—leading directly to the event horizon—may be only a few light-months across. Radio telescopes also detect huge jets emerging from the top and bottom of this system and reaching out for more than thirty thousand light-years on each side.

that also encompasses Karl Schwarzschild's original solution for a motionless object. Any spinning mass will tug at spacetime. Even the Earth does this, but to an extent that is extremely difficult to detect. However, things get pretty interesting when it comes to a black hole and the enormous stress it places on spacetime around its compact mass. In particular, because of light's finite speed, there is a distance away from a rapidly spinning black hole at which photons traveling counter to the twister-like spacetime could actually appear to stand still. This critical point is farther out than the distance we call the event horizon, from which no particles of light or matter can escape.

With all this in mind, a spinning black hole actually has *two* locations, or mathematical boundaries, around it that are important to know about. The outermost is this "static" surface where light can be held in apparent suspension, motionless. It's the last hope for anything to resist being swept around and around by the spacetime twister. Then the surface inward from that is our more familiar event horizon. Between these two surfaces is a maelstrom of rotating spacetime. It is still possible to escape from this zone, but you cannot avoid being moved around the black hole, since spacetime itself is being pulled around like a thick carpet beneath your feet. This rather spooky region is known as the *ergosphere* from the Latin *ergon*, which means "work" or "energy." Furthermore, neither the outer surface of this ergosphere nor the inner event horizon is spherical. Just like those of a balloon full of liquid, the horizons and surfaces around a spinning black hole bulge out toward their equators, forming what is known as an oblate spheroid.

Spinning black holes open up a bag of mathematical wonders. Most of these don't concern us for the purposes of our quest to understand the far-reaching effects of matter consumption, but they're fascinating and lead to some of the most outrageous concepts in physics. For example, the true inner singularity in a spinning black hole—that central point of infinite density—is not point-like at all, but rather smears into the shape of a ring. Not all routes inward arrive directly at this singularity, and objects may miss this bizarre structure altogether. Wormholes through to other universes and time travel are tantalizing possibilities in some cases, although the very presence of foreign matter or energy seems to thwart these hypothetical phenomena. It is intoxicating and magical stuff, but the most important piece that's relevant to our present story is that there is in fact a maximum rate at which a black hole can spin.

In that sense, black holes are remarkably similar to everything else in the universe. At a high enough rate of spin, the event horizon would be torn apart, and the true singularity would be exposed and naked. That's not a good thing for our theories of physics. Singularities are best kept hidden behind event horizons. If they weren't, then, in technical terms, all hell would break loose. Luckily, nature seems to prevent black holes from ever getting past this point, although, as we'll see, they get awfully close. In the 1980s the physicist Werner Israel demonstrated that the universe must conspire to stop a black hole from ever gaining maximum spin. Once a black hole has reached close to the highest rate of rotation, it becomes effectively impossible for incoming material to speed it up any more. Matter quite literally cannot get close enough through the centrifugal effect of the spinning ergosphere. This means that any further interaction with the external universe will typically act to slow down, not speed up, a maximally spinning black hole. In this way it is kept from tearing apart. Perhaps not surprisingly, this limit to spin occurs when the rotational velocity close to the event horizon approaches the velocity of light.

This brings us back to the English physicist and mathematician Roger Penrose's marvelous insight in 1969 that the rotational energy of a black hole can be tapped into via the surrounding spacetime twister. This mechanism is important because the accretion disk of material surrounding an eating black hole continues all the way into the ergosphere. It's perfectly fine for it to do so—it's still outside the event horizon. Within this zone, the relentlessly dragging spacetime will force the disk to align itself with the equatorial plane of the spinning hole. The same kind of frictional forces that allow the matter to shed energy will still be at play, and that energy can still escape the ergosphere. So matter in the disk continues to accrete through the ergosphere and inward to the event horizon. As the spinning black hole grows from eating this matter, it will also gain the spin, or angular momentum, of that material. Keeping all this in mind, we'd expect the most massive black holes in the universe to also be rotating the fastest, all the way up to the limit of maximal spin. This could be a terribly important factor in the next phenomenon we need to think about, which is all about siphoning off that spin.

› › ›

Jets of matter are a phenomenon we find in many situations here on Earth as well as out in the cosmos. We can start off by thinking about the jet of water that comes out of a hose. Water under pressure is confined in a tube, and when it emerges it has a tendency to just keep going in the same direction. The same principle holds elsewhere. For example, on a relatively small cosmic scale, as young stars gather up matter and become more and more compact, they too can propel flows or jets of material. These are impressive-looking structures when seen through a telescope. Particles of matter are accelerated out in northern and southern beams at velocities of about 60 miles a second. Eventually, they crash into tenuous interstellar gas and dust many light-years away, producing bright splashes of radiation. Supermassive black holes can produce jets of matter as well, but their nature is quite literally of a different order. Particles in this case travel outward at close to the speed of light—what is called an ultra-relativistic state. These are the extraordinarily fine and narrow lines or rays emanating from some galactic cores. They are also often associated with the rare, but impressive, radio-emitting dumbbell structures around galaxies that we encountered previously. Visually, we're tempted to think that the jets are somehow creating the dumbbells, but to be sure we have to better understand their origin and nature.

Just *how* jets of incredibly accelerated matter are formed is one of the most enduring problems of modern astrophysics—not, however, for want of ideas. Scientists have put forth a wide variety of possible mechanisms as contenders, many of which are at least superficially plausible matches to what we see in the universe. But the devil is in the details. Two basic things have to happen for nature to make a jet of matter. The first is that a physical process has to generate rapidly moving material. In the case of jets from black holes, these particles are streaking away at very close to the speed of light and seem to emanate from the poles of a spinning and spheroidal horizon. The second requirement is for this stream of ultra-high-velocity matter to be funneled into an incredibly narrow beam that can squirt out for tens of thousands of light-years. It's like a magical hose that forces all the water molecules to shoot out in near-perfect alignment so that you can accurately drench your neighbor at the far end of the street, if so inclined.

Funnily enough, there appear to be a variety of ways for nature to perform an extraordinary trick like this, and a big part of the challenge has been to figure out which mechanism is at play. For the extreme environments around a black hole, the answer seems to involve magnetism. When James Clerk Maxwell formulated his laws of electromagnetism back in the mid-1800s, he crystallized a description of how moving electrical charges, or currents, produce magnetic fields. These same rules apply to an accretion disk, the whirling hot plate of sauce around a black hole. A structure like this will be full of electrically charged matter. It's easy to imagine why it has to be. The temperature of its inner regions is so high that atoms are stripped of their electrons. Positively and negatively charged particles are racing around in orbit about the hole, and as a result, great currents of electricity are flowing. It seems inevitable that powerful magnetic fields will be produced, and as is their nature, they will extend away from or into the structures surrounding the black hole. As the material in the disk spins around and around it will pull these magnetic fields with it, but it will pull them most efficiently close to the disk itself, and less so above or below. It's not unlike taking a fork to a plate of spaghetti. The strands of pasta are the lines of magnetic field or force. The tip of your fork is like the sticky swirling disk of matter. Spin the fork into the spaghetti. The strands begin to wrap around, because the fork is pulling against the ones still lying on your plate. Above and below the disk around a black hole the strands of magnetic spaghetti are twisted into a funnel-like tube, leading away from both poles. It becomes a narrow neck of escape. Particles that boil off from the disk get swept up into these pipes of densely packed magnetic spaghetti and are accelerated even further as they spiral outward through and within this corkscrew. This should work incredibly well at producing a jet of matter. But to accelerate particles to close to the speed of light may need something still more. It may need a turbocharger.

When Roger Penrose demonstrated the principle of how rotational energy could be extracted from a black hole through the ergosphere, it may have seemed like an esoteric and immensely impractical idea to most of us. But there is another property of black holes that makes such energy extraction a very real possibility, and further supports Penrose's original idea. Scientists now think that a black hole can behave like an electrical conductor, which is an utterly counterintuitive idea in that the event horizon is supposed to hide all information from us. Indeed, only the mass and the spin of a hole are manifest through their effect on the curvature of the surrounding spacetime. At first glance there doesn't seem to be a way to paint any more colors onto these objects, to give them any more

Figure 12. A sketch of one way that a narrow jet of matter may be created by a spinning black hole. Magnetic field lines ("spaghetti strands") that are anchored in the disk of accreting matter around the hole tend to twist and wind up, creating a tube-like system that "pinches" gas and particles into a jet as they race outward.

properties. Yet there is one more piece of trickery that can occur because of the incredible distortion of spacetime just outside the event horizon.

Imagine you have in your possession an electrically charged object, such as a single electron. You can tell that it's electrically charged because if you move another electrically charged object around it, you can feel a force between the two. Like charges repel, and opposite charges attract. That force is transmitted through spacetime by photons, and it is all part and parcel of electromagnetic radiation. Now, let's say I'm going to whisk that electron away, place it just outside the event horizon of a black hole, and ask you to come along and look for it by sensing the electric field. Most likely, you're going to get somewhat confused, because the extremely curved spacetime at the horizon can bend the paths of photons, and hence of electrical forces, completely around itself. Even if the electron is placed on the opposite side of the hole from where you are, its electrical field will be bent around to your side. It doesn't matter what direction you approach the black hole—you'll still feel the electric force of the electron. It is as if the electrical charge has been smeared across the entire event horizon. The hugely distorted spacetime is creating an electrical mirage, except it is better than a mirage. It is equivalent to the black hole having acquired an electrical charge.

This is exactly the way that an electrical conductor behaves—say, a piece of copper wire, or a chunk of gold ingot. An electrical charge on these materials exists only on their surfaces. The truly remarkable consequence is that a spinning black hole, surrounded by magnetic fields, produces a difference in electrical potential, or voltage, between its poles and the regions toward its equator. The physicists Roger Blandford and Roman Znajek first demonstrated the idea that a black hole can do this in 1977. A spinning hole will quite literally become a giant battery. But unlike the little battery cells you put in a flashlight or a camera, where there is a one- or two-volt difference between the "+" and the "–", a spinning supermassive black hole can produce a pole-to-equator difference of a thousand trillion volts. Surrounded by hot and electrically charged gas from the accretion disk, this voltage difference can propel enormous currents. Particles are accelerated to relativistic energies and funneled up and away through the twisted magnetic tubes above and below the black hole. This is driven by the enormous store of rotational energy in the black hole. Theoretical calculations show that this alone can produce an output equivalent to the radiation of more than a hundred billion Suns. It may still be that more than one mechanism is at play across the universe for producing accelerated jets of matter, but this one is a leading contender for black holes. It also means that when we see a jet, we are seeing a signpost to a charged and fast-spinning black hole.

These jets of particles are relentless. They drill outward as they climb away from the black hole, and there is little in a galaxy that can stop them. They simply bore their way out through the gas and dust within the system and carry on into the universe. Intergalactic space is not entirely empty, however. Although incredibly sparse, atoms and molecules still exist out in the void, and over thousands of light-years the particles in the jet collide with these rare bits of matter. As a result, the very leading end of a jet sweeps up this material before it like someone hosing dirt off the sidewalk. But this intergalactic gas and dust cannot move as fast as the ultra-relativistic particles squirted out by the black hole, and eventually there is a cosmic pile-up of speeding matter. This train wreck of material builds into an intense spot where the jet particles are bounced, reflected, and diverted from their straight paths. It's not unlike shooting a hose at a hanging bedsheet: it gives a little, but mostly the water sprays out to the sides and back at you.

The deflected jet particles are still extraordinarily "hot," moving at close to the speed of light. Now they start to fill up space, still pushing other matter aside and outward into a shell- or cocoon-like structure that encompasses the jets, the galaxy, and the black hole. This is precisely what creates the enormous radio-emitting dumbbells extending for thousands of light-years around certain galaxies. The radio emission is coming directly from the jet particles themselves, as they cool off over tens of millions of years. How this cooling works is part of a fundamental physical mechanism in nature that was actually first discovered here on Earth, and almost by accident.

Since the late 1920s physicists have been studying the most basic subatomic building blocks of matter in particle accelerators. The idea behind these devices is simple in essence, and harks back to the earliest experiments with electricity and magnetism. A particle like an electron has an electrical charge, and so we can use electric and magnetic fields to move it around. We can then propel or accelerate it to extremely high speeds. As the particle gets closer and closer to the speed of light, all the wonderful effects of relativity come into play. Physicists have learned to exploit this and use the terrific energy carried by an accelerated particle to smash and crash into other particles, converting energy into new forms of matter and making the apparatus a microscope of the subatomic.

The exotic new particles generated in these experiments can be extremely unstable. For example, one of the simplest and most readily produced is the particle called a muon, sometimes described as a heavy electron. The muon is also electrically charged, but it is not stable and has a half-life of existence of about two microseconds before it turns into an electron, a neutrino, and an antineutrino. If you want to study the muon, you'd better be pretty quick on your feet. But if you accelerate a muon to close to the speed of light, you can give yourself all the time you need. The muon's clock will appear to slow down, and its brief lifetime can be extended to seconds, to minutes, and even longer. All you have to do is keep it moving fast. One of the ways to do this is to propel particles around and around a circular loop of magnets and electrical fields. The Large Hadron Collider and many of the other major particle accelerators in the world follow this design. It's a great solution for keeping your subatomic pieces under control. The problem is that a constant force must be applied to the particles to keep them flying around in a circle. When this force is applied using magnetic fields, for example, then in order to change direction the particles will try to dispose of some of their energy. This streams out as photons, and that happens even when the particles are not moving particularly fast. But when they're barreling around at close to the speed of light, a whole new regime opens up.

In the late 1940s, a group of researchers at General Electric in Schenectady, New York, were experimenting with a small device called a synchrotron, a cleverly designed circular particle accelerator. (In order to push particles to higher and higher velocities, the synchrotron tunes its electric and magnetic fields to "chase" them around and around. It's like a wave machine for subatomic surfers. It sends a perfect ripple of electromagnetic force around the track to constantly propel the particles and keep them zipping around a circular path. It synchronizes with them, just as its name implies.) The GE physicists were pushing their synchrotron to the limit to test its abilities. The experiment used an eight-ton electromagnet surrounding a circular glass tube about three feet in diameter. By cranking up the power, the scientists were pushing electrons in the tube to velocities close to 98 percent that of light, hoping to probe deeper and deeper into the atomic nuclei of matter

One afternoon, a technician reported an intense blue-white spot of light pouring out of one side of the glass vacuum tube just as they reached peak power. Surprised by this, the scientists fired up the accelerator once more, and again, at the highest power, it lit up a brilliant spot of light. They had inadvertently discovered a very special type of radiation predicted just a year earlier by two Russian physicists. The excited scientists at GE quickly realized what they were seeing, and since the phenomenon had previously been only a theory with no agreed-upon name, they christened it with the practical but rather unimaginative label of "synchrotron radiation."

They had discovered that when charged particles moving close to the speed of light spiral around magnetic fields and are accelerated in a sideways direction, they pump out radiation with very special properties. This is a distinct "relativistic" version of the energy loss experienced by any charged particle getting buffeted by magnetic forces. Remarkably, from this experiment in the 1940s comes the key to appreciating how the beams of matter from black holes cool off over cosmic time. In these splashing jets, the energy of motion in particles like electrons and the single protons of hydrogen nuclei is being converted into natural synchrotron radiation. It runs the gamut from radio frequencies to optical light and higher and higher energies like X-rays. It also comes with some quite unique characteristics. The ultra-high velocity of a synchrotron radiation–emitting particle results in the radiation pouring out as a tightly constrained beam in the direction it's moving in, just like the spot of light from the GE experiment. If you were standing off to the side you would not see anything. Stand in the path of the beam, though, and you'd be scorched by the intense radiation. Out in the universe this property is very clearly manifest. Jets from supermassive black holes are quite difficult to see from the side—they are thin and faint. But once the jet particles splash into the growing cocoon around a galaxy, their synchrotron radiation lights up in all directions: the glow of the dragon's breath.

So now we've arrived at a pretty good description of the ways in which our black hole monsters consume matter and belch their energy into the cosmos. Gas, dust, and even stars and planets that are swept into the accretion disk of a black hole can be torn apart by gravitational tides and friction-heated to very high temperatures. This heat causes the disk alone to glow with the power of many galaxies. The quasars are the most powerful examples of this, and they represent a bird's-eye view into the center of a disk surrounding a black hole. They are also extraordinarily efficient, eating just a few times the mass of our Sun per year in raw cosmic material. The spacetime twister of spinning black holes cranks up this phenomenon to a new setting on the amplifier, and it also gives rise to another energy outlet: ultra-relativistic jets of matter that streak across thousands, sometimes millions of light-years. We think that spinning,

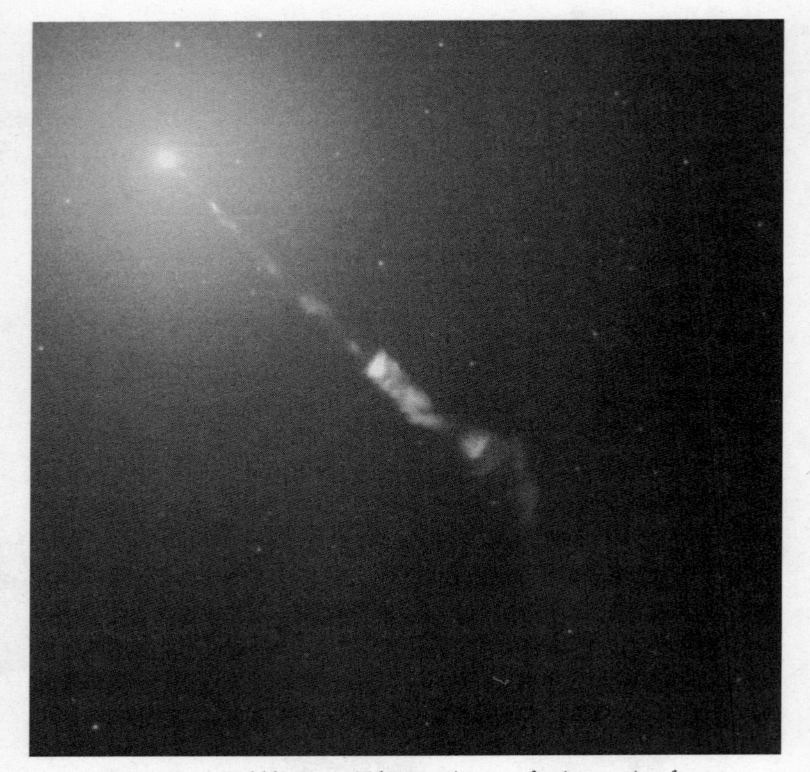

Figure 13. A Hubble Space Telescope image of a jet coming from the center of the galaxy called M87. This is a giant elliptical galaxy 54 million light-years from us. Amid the dandelion-like haze of hundreds of billions of stars, the jet extends outward more than five thousand light-years, glowing in blue-tinged visible light that is the synchrotron radiation of electrons moving at close to the speed of light. The black hole producing this jet is 7 billion times more massive than our Sun and is eating about a Sun's worth of matter every year.

electrically charged holes may be required to launch these sprays across the cosmos, and when they splatter into the intergalactic grasslands, their careening particles push aside great cocoons, glowing hot with synchrotron radiation. In this way a black hole that would actually fit inside the orbit of Neptune can produce these potent structures that extend over a hundred thousand light-years. That is as if a microscopic bacterium suddenly squirted out enough energy to inflate a balloon more than a mile wide. The monster is tiny, but its breath is enormous. The next challenge is to begin to investigate what this particularly virulent exhalation does to the universe. But before that it is worth pausing for a brief recap—and to consider again the nature of what we're dealing with.

› › ›

Black holes really are like something out of a fairy tale. The great American physicist Kip Thorne, who has played a central role in the development of black hole theory and the quest to find these objects, puts it nicely: "Of all the conceptions of the human mind, from unicorns to gargoyles to the hydrogen bomb, the most fantastic, perhaps, is the black hole . . ." In my brief version the story of these massive monsters began with the nature of light—something so commonplace, seemingly mundane, and part of our everyday existence. Yet the reality of light is actually quite fantastical. Here is a phenomenon that can be described in terms of electric and magnetic forms that behave both like waves and then as particles, moving through the vacuum of the universe like a snaking rope made of sand. Not only that, but it is light's constant pace that actually *defines* what we mean by space and time. Furthermore, the properties of matter that we call mass and energy do something extraordinary: they influence the very essence of this spacetime. They distort it, curve it, warp it. Reality is twisted and bent to make paths that we cannot comprehend with our biological senses but that we are literally compelled to follow as we move through space. Out in the universe it is these paths that underlie the vast neuronal forms of the cosmic web of matter as it coalesces and condenses into structures. Those structures fragment and flow into smaller structures. Eventually, because of the particular balance of forces and phenomena in this universe, matter can accumulate and concentrate to such an extent that it seals itself away from the outside.

Primal creatures are born in this process. Young and ancient black holes are the magical boxes that gobble up unwary passersby. Their event horizons are like punctures in spacetime, places that drain all the colorful and complex beauty of the cosmos out of sight. In a different universe, with different rules, this might happen quietly and discreetly. In this universe, our universe, it's usually a painful and ferocious process. We now know that matter does not go gently into the night. And like beasts grown out of other beasts, the black holes we find at the centers of galaxies have become monsters that sit inside their great castles. Their sheer size allows them to consume enough matter with enough violence that they light up the cosmos like flares tossed to the roadside. These monsters are a long way away and they've been around almost forever, a fascinating fact of life but one that we might at first assume to be unimportant to us. Yet in ancient fairy tales and myths, giants helped carve the world into its present form and provided the landscape we enjoy. Now they lie dormant, except for the rare occasions when something stirs them back to life. Perhaps we need to consider if this isn't also true of those real-life giants out in the cosmos.

Our investigation into this question through the history and life cycle of black holes is vibrant, and it continues as scientists race to new theories and observations. Many of us find it particularly intriguing because of the interplay between so many strands of scientific inquiry. In many respects that has always been the hallmark of black hole science. Both relativity and quantum mechanics were necessary to explain how black holes could actually come into existence, and astronomy operating at multiple parts of the electromagnetic spectrum is necessary to find the signposts to real black holes out in the universe. Although currently neither the physics of accretion disks nor that of astrophysical jets is complete, there may be deep connections between the microscopic scales that help determine things like friction in accretion disks and the vast scales of cosmic structure. It may be that there will be a "Eureka!" moment when we finally understand precisely what happens in these environments. It may also be that the physics is just too complex and variable between different instances, and a single crystal-clear description will elude us.

These challenges already tell us that black holes can be very messy eaters. But oh, what eaters they are! Whether or not we can pin down their precise table manners, we can most definitely see the consequences of what they do to the universe around them. It is the story of those consequences that will reveal some of the deepest and most puzzling characteristics of the universe that we have yet encountered.

5

BUBBLES

Carl Sagan once said that to make an apple pie from scratch you must first invent the universe. He was right. And in inventing the universe you will need to build all the objects and structures that we find in it. These are the planets, stars, white dwarfs, neutron stars, black holes, gas, dust, galaxies, galaxy clusters, and superclusters. Eventually, when this cosmic mix has cooked for long enough, the molecular arrangements will emerge to produce that apple pie. But how does the universe actually build all this stuff? It's a question for the ages. We have always wondered how our surroundings came to be. Perhaps we've sat around our fires and shelters asking one another about the looming silhouette of a great mountain against the twinkling stars, or the brightly lit disk of the Moon. Where did *that* all come from? For that matter, where and how did we spring out of these monumental forms?

The origin and evolution of objects and structures in the universe is a central and critical question for modern astrophysics, and it is arguably one of the biggest unfinished puzzles in science. In truth, one reason it is not yet complete is because it is a hugely complex problem that stretches both our physics and our imaginations to the limit. We may come up with clean and elegant fundamental rules for how the physics of the universe works, but nature's application of them is often extraordinarily messy. That is also, of course, part of the fun. For us it's also a key issue for dealing with the effects that black holes have on the universe, and we really need to get an understanding of the cosmic laboratory in which they are at play. To do this, we can split a big problem into simpler parts to understand it better. In this case, the growth of cosmic objects naturally divides into two large pieces. One is about construction; the other is about preventing that.

Let's deal with construction first. The glue that the universe builds cosmic structures with is gravity. Gravity springs from Einstein's mathematical framework for how mass distorts our stiff but flexible spacetime to create its own future paths. The number and variety of objects we see in the universe is in part determined by the effect of gravity on the tiny bumps and kinks of matter that we started out with almost 14 billion years ago. Had the baby universe been perfectly smooth and uniform, it would have remained dull and boring. With no seeds of structure there can be no growth. Exactly how many bumpy and kinky seedlings there were in the very young universe, and where they came from, is part of another fascinating story. For now it's enough to say that we think we have a pretty good idea of what they looked like. We also feel pretty confident that most of the matter in the universe is dark. This soup of ghostly but massive particles is distinct from the kind of stuff that stars, planets, and human beings are made of. It is the combination of dark and normal matter coalescing and moving around owing to gravity's glue that provides a very big piece of the building plan for our universe.

We also know that eventually the continuing expansion of spacetime following the Big Bang will put an end to any construction. The immense stretching out of space will isolate material into islands. More and more, matter will end up in small dense objects or dispersed farther and farther apart until nothing new is formed. That, however, is very far in the future.

The second piece to the problem of where all structures and objects come from is a little trickier: it's all about resistance to building, or even destruction. So is there a yin to the yang of cosmic evolution? Indeed there is. Matter in the universe, rather perversely, also creates many obstacles to its own assembly. The most fundamental hurdle for matter comes from the basic phenomenon of its own pressure, which is in turn related to its temperature. The component atoms or molecules of a material such as a gas are buzzing and bustling around, and we gauge this by talking about how hot or cold the gas is. The hotter the gas and the faster the typical motion of these particles, the greater the particles' thermal energy. The colder it is, the more sluggish the particles get. Eventually, close to absolute zero, they should all stand still, but for their inherent quantum wiggling and jostling.

What we experience as gas pressure here on Earth is the combination of this thermal motion and the number of atoms or molecules in a particular region. With a ping, ping, ping, the gas particles of our atmosphere bounce against our skin, inside our lungs, and against one another. When you blow up a balloon you are filling it with trillions of air molecules that beat against the rubbery material, causing it to stretch and expand. The thermal motion of the gas creates this property of matter called pressure, which resists efforts to contain it. This is precisely how pressure and temperature work against gravity. Matter will try to fall into, pour into, the deep wells and bowl-like distortions in spacetime caused by mass. But the incessant motion of that matter is like having an infestation of springy fleas that you're trying to trap inside smaller and smaller boxes. The moving molecules just don't want to be confined. It's further complicated because matter tends to get even hotter as it becomes compressed. The same thing happens when you try to pump air into a bicycle tire. The forces of gas pressure resist compression, and some of the energy from your arms is converted into heat, causing the pump to get warm. That heat comes from the speeded-up thermal motion of the gas particles. The hotter the gas gets, the greater the pressure. It is a major obstacle to building objects in the cosmos, yet evidently not an insurmountable one, or you and I would not be here discussing it.

Objects can also explode. Massive stars have a bothersome tendency to end millions of years of nuclear fusion in great cataclysms we call supernovae. Similarly, white dwarfs may be fed just a little too much matter and exceed the critical Chandrasekhar mass, the largest-size object that can be supported by the quantum electron pressure that we encountered before. They implode disastrously. Radiation and particles burst forth to disperse all that carefully gathered material, like an impatient child getting frustrated with a house of cards. And we've seen that black holes, small and large, can generate huge outputs of energy with incredible efficiency. All these phenomena act against the gravitational gathering of matter, but despite this our universe clearly reaches a balance. It has to, or else there would either be nothing much present except tenuous gas and dark matter—or all matter would be locked up in black holes. The success of constant construction is clearly tempered by the success of ongoing obstruction; we are surrounded by a shifting and dynamic impasse. And a key discovery that will help us understand this impasse, this point of equilibrium, begins with an extremely famous and rather intimidating family.

› › ›

First there was old grandpa Erasmus Darwin, a physician and a highly regarded and historically important natural philosopher in the 1700s. Then, two generations later, along comes Charles. He sails off from England in his twenties for a five-year journey to the exotic southern oceans of Earth. He returns and later helps revolutionize our view of the nature of life. One of his sons, George, makes profound contributions to physics and celestial mechanics. And later one of George's sons, named Charles in cyclical fashion, becomes a highly respected physicist who helps apply quantum mechanics to the problems of atomic physics in the early twentieth century. Pity the Darwins' neighbors—personally, I'd focus on making my lawn look better than theirs.

In this brilliant family, it is Charles's son George Darwin who plays a small but pivotal role in advancing our understanding of how matter in the universe ends up making planets, stars, and even the great clusters of galaxies. It begins with an almost offhand, but deeply insightful, comment by George in the late 1800s. At the time, scientists were working to understand the origins of stellar systems in our own galaxy. A popular theory was the "nebular hypothesis," which posited that stars and planets formed out of interstellar gas and dust, somehow coalescing and condensing out of this material—although how and why an interstellar cloud, or nebula, would want to do that was a point of uncertainty. While George Darwin's main scientific efforts went into the complex subject of gravitational tides on planets and moons, he was also very aware that the nebular hypothesis really needed someone to tackle the physics behind it. In a paper published in 1888, he succinctly described what was missing from the theories of the time: the mathematical description of how a rotating cloud of gas could give way to its own gravitational forces to condense into stars and planets. It was a clearly phrased challenge, waiting for someone bold enough to pick it up.

More than a decade passed, and finally, in 1902, a young physicist employed at the University of Cambridge named James Jeans took George's comment to heart. Consulting with the now-senior Darwin, he dutifully produced a fifty-three-page treatise, "The Stability of a Spherical Nebula." In this work, Jeans provided the mathematical and physical basis for understanding the fundamentals of what we now call "gravitational collapse." The essence of it is simple, the practice rather more complex. Jeans established that in a structure like a nebula there are two opposing forces at play. One is gravity. The matter in a nebula has mass, and so it will tend to fall together, shrinking itself. The other force is the natural pressure of the gas. This is the springy force that resists the inward fall of material.

A blob of dense nebula represents more mass than a blob of less-dense nebula, so the greater the density, the greater the gravitational forces at work. But density is also related to gas pressure, temperature, and composition. Jeans saw that high temperature and low density would make it harder for a nebula to condense to make objects like stars. Conversely, low temperatures and high densities would make it easier for gravity to pull material together. Jeans also realized that if you measured just the temperature and density in a nebula, you could immediately calculate the size of a region that would be hovering in balance, just poised to collapse. A smaller region would have insufficient gravity to overcome its gas pressure. A bigger region would have insufficient gas pressure to resist gravity's embrace. This critical point later became known as the Jeans Mass.

In other words, if you find a nebula that is bigger than its Jeans Mass, then it is almost inevitable that it will be in the process of collapsing and condensing to make stars. Similarly, any cloud of gas that is actively cooling down by emitting radiation stands a good chance of cooling enough that it begins to collapse under its own gravity—especially if its mass is only a little less than the value of the Jeans Mass.

But surely one can just *see* whether or not a nebula is collapsing, right? Why bother with all this calculation? The problem concerns the human timescale versus the timescales of the cosmos. We'd have to wait around for hundreds of thousands of years to really notice a nebula collapsing to make stars. We're just too puny and short-lived. Instead we must rely on clues like those we get from Jeans's equation. He found a way for us to deduce actions of matter that are happening at a snail's pace from our terrestrial perspective.

We now know that there are many hideous complications to this simple picture. These include the elastic-like effects of interstellar magnetic fields, flowing motions in the nebula, and the endless lumpiness and complexity of material spread around in our galaxy. However, Jeans's insight is still critical. In a general form it applies across the universe, from the very first generations of stars to those forming in the spectacular Orion nebula in our night sky. Gravity must always overwhelm pressure in order to make objects in the cosmos. It also provides the basis for the next piece of our story, which is all about places not behaving the way you might expect them to.

›››

Clusters of galaxies might not at first seem to be the likeliest candidates for unveiling the mysteries surrounding the life cycle of black holes in the universe. While a supermassive black hole can occupy a volume similar to that encompassed by the orbit of Neptune, a big cluster of galaxies can occupy a region some 30 million light-years across. The black hole is only 0.00000000001 times the size of the cluster. That's the size of the period at the end of this sentence compared to one-third of the distance to the Moon. Nonetheless, there is indeed a very special relationship between these two vastly different structures, and it's one that is connected to the constructive and destructive elements of the cosmos.

I've said it before: galaxy clusters are the cathedrals of the universe. These vast systems can contain hundreds, even thousands of galaxies. In this way clusters are the largest "objects" in the cosmos, the great big conglomerations of material at the intersections of the cosmic webbing of matter. As such, they also represent a nearly closed environment, an astrophysical biosphere in which physical phenomena are captured and contained. They are gravitationally bound together within the spacetime distortion of a quadrillion Suns' worth of mass, composed of dark matter, gas, and stars. As a result, the escape of material is seldom an option.

In these intergalactic biospheres, the majority of normal matter exists in the form of extremely hot and tenuous gas—gas that's so hot that electrons are stripped from atoms to leave them as ions, the positively charged nuclei and negatively charged electrons coexisting to create plasma. This plasma outweighs all the stars in the galaxy clusters. Most of it is primordial hydrogen and helium, slurped down into the gravity well of the cluster by the same circumstances of imbalance discovered by James Jeans. This deep well is in turn dominated by the unseen dark matter that outweighs all the normal matter in gas and galaxies by about ten times. The captured gas falls ever inward within the gravity well, but as it accelerates it crashes into itself, and converts the energy of this waterfall-like pouring motion into the thermal motion of individual atoms. This is how the gas heats up, and a temperature of 50 million degrees is not unusual inside the biggest galaxy clusters. And the more massive the cluster, the higher the temperature can go.

Hotter gas means higher pressure, and that can stop gravity from compressing the gas any more. Instead it just sits and seethes in the gravitational bucket of the cluster. But over time this gas can also cool off. It can do this by rearranging any electrons that have managed to reattach themselves to ions, squirting out photons of light and releasing that energy. It can also cool as the electrons are decelerated by the electric fields between themselves and the oppositely charged ions of the gas.

This is much like the shrieks of rubber hitting rubber at the bumper-car rink in an amusement park. That noise is energy being lost as the cars whack into their surroundings. Similarly, electrons buffeted about in a plasma emit photons of light to bleed off energy. It's like the processes of radiation emission seen in particle accelerators that we talked about in the last chapter. The scientific name for the phenomenon is a wonderful mouthful: *bremsstrahlung* (brems-*stra*-lung) comes from the German *bremsen* (to brake) and *Strahlung* (radiation), and literally means "braking radiation." Among its many fascinating characteristics, bremsstrahlung from the cooling gas inside galaxy clusters is not visible to the human eye because it's in the form of X-rays. And, like the bumper cars, the more tightly packed together the electrons are, the more energy they can get rid of—the more rubbery shrieks of X-ray photons—and the faster things cool down.

The first evidence for the existence of this superhot gas in clusters emerged in the late 1960s during the dawn of X-ray astronomy. Unlike the sharp, pinpoint-like X-ray emission of neutron stars and black holes, galaxy clusters are big and cloudy. When you see the X-ray image of gas in a galaxy cluster, you are a direct witness to the amazing dent made in spacetime, filled up with matter like water poured into a bowl. Yet this gas is remarkably tenuous. A cubic meter may contain a total of only a thousand ions and electrons. We only notice it because we see the cumulative light from a depth of millions of light-years into this thin fog. This gas also squeezes in toward the core of a cluster, because of the typical shape of the spacetime bowl. It's a bit like a soufflé that's partially successful. All is light and fluffy, except for the embarrassingly thick gummy patch at the bottom. This creates an intriguing conundrum: We know that this bowl of gas cools off by emitting X-ray photons. The primary route by which it does that (the bremsstrahlung) is related directly to how dense the gas is—how many of the electrically charged electrons and ions there are in any given region. If particles are closely packed, the cooling happens faster. So the cluster should be cooling most rapidly at its center, where the gas is densest. But cooler gas means lower pressure, which leads to gravity squashing matter further together, making it even denser and allowing it to cool even faster.

This can act as a runaway process, not unlike a car perched on a hilltop without its emergency brake on. At the top of the hill the slope is gentle, and if I inadvertently lean on the car, it moves just a bit at first. But if I fail to jump in and apply the brakes, then it picks up even more speed, until eventually all I can do is watch in horror as it whizzes down the hill and off an inconveniently located cliff.

Essentially the same thing *should* happen in a galaxy cluster. The thicker gas in the core cools faster by pumping out more X-ray photons. As it cools its pressure will decrease, and gravity will cause it to slump inward. If the temperature of this gas drops by just a factor of three, it will shrink down inside the gravity well to become twenty times denser. It's James Jeans's argument for the collapse of nebula gas all over again: when temperature and density drop below a magic level, gravity takes over. Inside a galaxy cluster, this rapidly cooling gas will begin to roll down the hill, toward the center. But unlike my unfortunate car trundling down a slope, this gas is also supporting lots more gas above it, the rest of the soufflé. Take away that support and the material above will also slump inward. So the outer gas will in turn pour down the bowl sides, increasing in density and cooling more rapidly. It's as if I had tied my car to the front of a whole chain of other cars, all destined for the cliff edge.

In the outer realms of a galaxy cluster the gas cools at a very slow rate. In the center, though, the runaway process can cool down hundreds of times the mass of the Sun in gas *every year*. That may not sound like much, but a typical cluster has been around for billions of years, so that adds up to an awful lot of material turning into a thick cold nebula. And thick cold nebulae, as James Jeans saw, have a tendency to collapse further, condensing into stars.

This characteristic of galaxy clusters began to emerge into scientific discussions in the mid-1970s. By this time, early generations of orbiting telescopes had found intriguing signs of dense and bright X-ray emitting gas in the very centers of a couple of these huge systems. One of the scientists trying to understand these measurements was Andrew Fabian. The English-born Fabian was one of a new breed of astronomer, cutting his teeth as a doctoral student with rocket-borne X-ray detection experiments. Launched from Australia and Sardinia, the rockets gave him ten-minute-long peeks above Earth's atmosphere and into the cosmos. Continuing his postdoctoral career at the University of Cambridge—still his scientific home today—Fabian, together with his student Paul Nulsen, joined a few other scientists around the world in eagerly studying the physics behind the intriguing new images of galaxy clusters. At the center of some, the density of the gas and its X-ray emission indicated it was cooling down in less than 10 million years, the blink of a cosmic eye. The investigators quickly realized that this could be precisely the signature of a runaway process, and this great cosmic downfall of gas soon earned the name "cooling flow."

Most large galaxy clusters also have a big galaxy sitting in their centers. These central objects are of the elliptical class, each a dense dandelion-like cloud of hundreds of billions of stars. All that cooling cluster gas should end up in these central objects. It could even be responsible for building these galaxies in the first place, and they should still be rife with the formation of new stars. But there's a catch: something is amiss. Nature is not playing ball. Gas cools in clusters, all right, but most of it never actually gets to the point of making stars—a huge problem for what scientists thought was an obvious process.

By the early 1990s, X-ray observations were sophisticated enough to allow for more precise estimates of just what was going on in the centers of the clusters. In most of these systems, the central concentration of X-ray light appeared to match up with the theoretical predictions for "cooling flows." In some cases the X-ray data was good enough to allow astronomers to estimate the gas temperature itself, which indeed seemed to be dropping in cluster cores. Many scientists staunchly advocated that cooling flows were vital parts of galaxy clusters. They had good reason to. Everything seemed to fit together—or at least almost everything.

The biggest stumbling block was the sheer amount of material that was thought to be cooling down until it no longer emitted X-rays. If it was turning into cool nebula-like clouds these should in turn be condensing into new stars, lots and lots of them. But there wasn't much evidence to support this. The giant galaxies at the centers of clusters simply didn't contain a huge excess of young stars. Where cooler gas *could* be seen, there was very little compared to expectations. In 1994, Fabian tried to come up with a number of explanations for what was going on. It was conceivable that whatever the cooling gas was turning into was simply dark, and effectively invisible. It could be dust, or small and tepid stars. Some gas could be dropping into near invisibility as cold molecules of simple compounds like carbon monoxide. It was also possible that more complex physics was controlling the gas and hiding it. Magnetic fields might be channeling and constraining the gas, and perhaps hot and cold gases were coexisting in complicated structures that fooled our telescopes.

As with so many puzzles in science, new observations and new data would make all the difference. In 1999, and within five months of each other, two mammoth telescopes launched into orbit around the Earth, and they would prove to transform our view of the universe. One was NASA's Chandra X-ray Observatory, and the other was the Newton Observatory of the European Space Agency's X-ray Multi-Mirror Mission (or XMM-Newton). Both were designed to collect more X-ray light from astrophysical objects than ever before, and to make the most detailed images and spectra of that light. With this new precision, scientists could now monitor the X-rays from galaxy clusters well enough to closely track the behavior of the cooling gas by exploiting another of its unique characteristics.

In among the hydrogen and helium of this gas are the same kinds of elemental pollutants that we find everywhere in the universe. Produced by generation upon generation of stars and wafted, blasted, and blown out of individual galaxies and protogalaxies, these heavier elements permeate the denser regions of the cosmic webbing. Some of them are particularly visible in X-ray light from hot gas. Iron, for example, has a complex energy hierarchy for the electrons held around its atomic nucleus. Oxygen, the most abundant element in the universe heavier than helium, has similar properties. In these cases, even at temperatures of millions of degrees electrons can stay engaged with their individual atoms, and energetic X-ray photons are absorbed and released at very specific wavelengths or "colors." The highly advanced instruments on board Chandra and XMM-Newton could sniff out the X-ray photons from these heavy elements. This light provides a unique fingerprint directly related to the gas temperature. It's like having a thermometer inside a galaxy cluster, and astronomers were quick to put these tools into action, swinging the great observatories to gather up light from the brightest systems in our nearby universe.

Here was the gas, cooling. Down and down it went. And then . . . nothing. You can imagine the scientists' consternation. Suspecting a mistake, they quickly reanalyzed the data. But there was no mistake. In cluster after cluster, the gas cooled down as expected, and then, just as it reached a temperature of a little more than 10 million degrees, it stopped. Not only did it stop, it didn't even accumulate at that minimum temperature—there was no vast snowdrift of piling-up material. You might expect a great mass of this gas to be growing as more and more poured in through the outer cooling flow, but it wasn't doing that. To all intents and purposes, it was mostly vanishing, with just a trickle carrying on down to lower and lower temperatures. It was as if the chain of cars rolling down the hill got a certain distance and then just conveniently disappeared. It felt like watching a great ocean liner gracefully sailing off to the horizon, and then suddenly turning into a dinghy before dropping out of sight.

Clearly, something was happening to all this gas. Perhaps an unknown mechanism was heating the gas in a targeted fashion, cooking up the cool stuff and getting it quickly back into the general mix before astronomers noticed it. Or perhaps, with the right arrangement of magnetic fields, thermal energy could be channeled to the cooling gas to warm it up—a great system of under-floor heating. The X-ray thermometer that astronomers were using relied on the precise mix of heavier elements like iron and oxygen with unpolluted gas. Fewer of those richer elements could skew the measurements at lower temperatures. Perhaps the just-cooled gas could be mixing up with either much hotter or much cooler gas, or could even be obscured by a hitherto unseen blanket of cool material that simply blocked the X-rays from our view.

There was another possibility. Maybe energy from a central supermassive black hole was halting the cooling. But how? The ability of a black hole to squirt out jets of particles, jets that could then splash out into great lobes of seething relativistic electrons and protons, was a tantalizing candidate. We knew that big central galaxies in clusters often harbored such structures. These were places where radio maps traced out colossal clouds and arcs of particles. Astronomers had certainly considered the impact these feeding black holes might have on the young galaxies and cooling gas we could see in the more distant universe. But something was needed in our neighborhood galaxy clusters: a clear signpost, something that could tell us exactly where to look. In the end there were several such signs, but one in particular is so big and clear that in retrospect it's almost embarrassing that we hadn't connected the dots before.

The answer begins with Perseus—not the monster slayer of Greek mythology, but a huge cluster of hundreds of galaxies, named for its sky placement in that hero's constellation but actually located about 250 million light-years from our solar system. The Perseus cluster is one of the largest such structures in our cosmic neighborhood, and the brightest such object in X-rays from our viewpoint. If we could see it with our eyes, it would cover a patch of sky four times broader than the full Moon. Adding up all the stars, gas, and dark matter in this great agglomeration, it totals a staggering 700 trillion times the mass of our Sun—a thousand times the mass of the Milky Way. This huge collection is spread over a three-dimensional region that stretches across 12 million light-years. As in all such vast gravitational wells, the bulk of the normal matter in Perseus consists of gas at extraordinary temperatures. Heated to tens of millions of degrees, it glows with X-ray photons. Just as in all the other great clusters, the gas is trying to cool off. And just as in so many clusters, within the center is the unmistakable signature of particles that have been spewed out from around a supermassive black hole, glowing with radio emission.

A puzzle about Perseus had emerged in the early 1980s with its first detailed X-ray images. Perseus was great to look at—it was big and bright, and its glowing gas could be seen spread across hundreds of thousands of light-years, brighter toward the center and getting ever fainter toward the edges. But there was a strange dark zone in one corner of this huge structure, an area that looks like a dirty thumbprint on these images, blotting out perhaps a hundred thousand light-years of X-ray glow. A decade later, new telescopes and instruments revealed the same thing. Perseus had a gap in it. In 1993 the German astronomer Hans Boehringer used the latest data to begin to finally unravel the mystery. Within the core region of Perseus were two other vast dark hole–like gaps inside the hot, glowing gas. When Boehringer and his colleagues placed a map of the radio emission from Perseus over their new X-ray image, the two prominent lobes of radio light lined up almost perfectly with these voids in the gas.

There had been growing suspicion over the years that the high-speed particles injected into a cluster from a central black hole jet would have to push aside the cluster gas. Not only that, but they would constitute a far lower-density material than even the tenuous cluster plasma. This suggested that the inflating lobes of radio-emitting particles should be buoyant. If so, then they should quite literally float in the cluster. But it was not yet clear whether this could really happen. Perhaps these lighter bubbles would simply dissipate and fizzle out. Along with the enormous cooling flow puzzle, this was a huge question to answer. It was evident that we needed to take a very, very careful look deep inside one of these systems.

Fabian and his colleagues set out to study the Perseus cluster in unprecedented detail, using the Chandra observatory. After some intriguing initial results, they decided to go for broke. They needed the most precise image of Perseus possible. Only this would allow them to peel apart the data to get at the gold nuggets inside. Over the course of two years, they accumulated the data they needed, until they had almost 280 hours' worth of photons—nearly a million seconds altogether—and from those they finally generated a new image of Perseus.

And what an image it was. With this incredible new visual fidelity, Perseus took on a whole new texture and flavor. It looked like a pond after a giant pebble has been thrown in, interrupting its smooth surface. There are clear bubble-like gaps, and there are *ripples*—actual *waves* through intergalactic space. All these are marching outward from the supermassive black hole at the core.

Fabian's team looked at the data every way they could, holding it up like a faceted diamond to see the shifting colors and projections from within. The mysterious gaps high up in the outer regions of the cluster were indeed rising bubbles. Whatever highly energized particles the black hole had squirted out millions of years ago to inflate these forms had long since cooled down, invisible even to sensitive radio telescopes. But the particles live on, holding the cluster gas at bay. These are ghostly cavities in Perseus's body, buoyant structures in a sea that is hundreds of thousands of light-years

Figure 14. An X-ray image of the hot gas in the inner regions of the Perseus galaxy cluster, showing the dark bubbles that have been blown by the black hole at the center and the great ripples of sound waves set in motion across the structure. At the very center are signs of some darker, cooler gas, looking like dirty strands of coagulated matter.

across. Down toward the core there are new bubbles, these still filled with hot electrons shedding radio-wave photons. The ripples between these floating bubbles are subtle, gentle structures. They are actual sound waves, the booming call of a leviathan. The time it takes for light to travel from one wave crest to the next is equivalent to the passage of all recorded human history.

Fabian himself is an avid rower. More often than not, if you're taking a morning stroll along the banks of the River Cam passing through Cambridge, you'll spot his boat skimming along the gently flowing water, its V-shaped ripples expanding outward, lapping softly at the riverbank. Watch him move upriver, and you see the same processes at work that are taking place in Perseus. As the boat pushes and displaces the fluid around it, some of that energy dissipates across the water, far away from its source. The lap and splash of the shoreline waves comes from energy that is generated within the muscle fibers of human arms and transferred across a river's surface. The supermassive black hole in a cluster core does the very same thing.

These structures in Perseus help us understand how the central supermassive black hole holds cooling matter at bay. The buoyantly rising bubbles can lift and push cooling gas aside, preventing it from funneling all the way down to the core. And like a vast musical organ, the bubbles set sound waves in motion that can disperse energy throughout Perseus, keeping it at a perfect simmer. We don't yet know if these rippling pressure waves, like the rolling thunderclaps of a distant storm, are entirely responsible for halting the flow of cooling gas into Perseus's inner core. But if we carefully measure these waves and compute how much energy they can push out across the cluster, it is certainly enough to balance out the energy that the cooling gas loses in X-rays.

A simple physics experiment brings the principle to life: place an open loudspeaker from a music system so that the speaker is on its back, forming a shallow cup. Put a sprinkling of sand or rice grains on the speaker. They roll and slither down to the middle. But then play music through the speaker with the volume turned up, perhaps a good bit of Bach or some heavy metal. The bass notes (or longer sound waves) vibrate the speaker and push at the grains. If you find the right pitch and volume, they'll spread back out up the sides of the cup, bouncing and agitated and unable to slump to the center, just like the gas in Perseus.

Every few million years a supermassive black hole in Perseus is being fed matter. When this happens, an outburst of energy from its jets and radiation inflates bubbles of high-pressure, fast-moving particles into the comparatively cooler and denser gas of the cluster. As these bubbles inflate, they act like the vibrating blast from a massive pipe organ, pushing off a rippling pressure wave that spreads out across the system. As it passes it releases energy into the gas, helping to prevent it from cooling and rushing inward. The black hole is driving the ultimate subwoofer, an audiophile's fantasy. The note it's playing? Fifty-seven octaves below B flat above middle C, in case you were curious. That's approximately 300,000 trillion times lower in frequency than the human voice. And the power output is a planet-disintegrating 10^{37} watts. Supermassive black holes can make you a very, very nice sound system.

We now know that Perseus is not the only place where this is happening: there is evidence for bubbles in 70 percent of all galaxy clusters. Here, too, the deep booming notes of these systems and their rippling undercurrents help moderate and regulate the cooling and inflow of gas. Here, as well as in isolated galaxies, we see evidence for a dynamic balance in the struggle between matter trying to build structures and the forces of disruption. In engineering this is called a "feedback loop." A simple example is a device used for centuries to regulate the speed of engines, from those driven by windmills on into the era of steam power. The centrifugal or "flyball" governor is a pair of opposing metal balls hanging like twin pendulums on stiff wires from a vertical rod. The rod is connected to the spinning axle of an engine. The faster the engine runs, the faster this vertical rod spins, and the higher the two metal balls rise as centrifugal forces push at them. But as they rise, the wires holding them can transmit that movement to a valve or throttle that slows down the engine and prevents it from running too quickly.

We think a similar thing happens with feeding black holes. In essence, the greater the amount of matter that falls toward the hole, the greater the energy output, and the harder it becomes for matter to reach the hole in the first place. This is the effect of the ripple-producing bubbles in a galaxy cluster. Although this serves to slow down the black hole gravity engine, it's rather sporadic and jerky. Unlike a beautifully smooth mechanical governor, a piece of high engineering from the Industrial Revolution, the fluid and fickle nature of gas and astrophysics results in something unique every time. Occasionally, a particularly dense patch of gas or stars manages to cool through to the core of a galaxy and descend into a black hole's grasp. The hole flares up brilliantly and squirts out a bubble-inflating jet for a million years before running out of fuel. At other times, there may be just enough gas to trigger a mild response.

In clusters where a trickle of gas manages to cool all the way down before being heated and rearranged, there are ethereal fingerprints. Astronomers' images in optical, ultraviolet, and infrared light reveal wispy threads and filaments of dense, warm material, cobweb-like structures of luminosity around the central zones of these clusters, no longer emitting X-rays. The cluster Perseus contains such forms within its central elliptical galaxy. Reaching across tens of thousands of light-years, they appear like strands of curdling milk in a great vat of matter. At the enormous distances of galaxy clusters it is impossible for us to see individual stars. They all merge into fog-like hazes. Nonetheless, by examining the spectral fingerprint of these milky filaments, we can tell that in among them are young large blue stars. The only viable explanation is that they are forming here, the end point in the journey of intergalactic gas that has settled its way down through the cluster. For now, escaping the petulant fist-waving of the supermassive black hole, it finishes as a warm nebula that then cools further to make new stellar systems. Eventually, these may be fodder for the next cycle of black hole activity, but not yet.

In the places where this happens, as much as a few times the mass of our Sun is converted from gas into stars every year. This rate provides a good match to the overall proportions of stars to gas in these central cluster galaxies. This is strong evidence that the *reason* we see the number of stars that we do in these galaxies is because the black holes are controlling the production line. The balance set by nature in these great systems of feedback is literally written in the stars. And that point of impasse stems from the fundamental nature of black hole physics, from electromagnetism to curved, spinning spacetime.

Here then, in the special environments of galaxy clusters, are clues to one of the major routes by which the universe builds new stars and galaxies. It is a startling display of how a supermassive black hole functions as a cosmic regulator, making sure the porridge of intergalactic matter is not too hot and not too cold. There is still much to learn about these mechanisms. There are indications that the supermassive black holes in the cores of clusters like Perseus may also be the fastest spinners. Faster-spinning holes are more efficient at producing energy. It makes sense: cluster-bound holes are the end recipients of a vast reservoir of intergalactic matter. If they weren't big and efficient at pushing back, we would see far more cooling gas converted into stars, and it would be a very different universe if that had been the case.

Massive black holes are also clearly present in the cores of other galaxies, including those that are not part of bigger clusters. The most powerful jets in these systems have little surrounding medium into which to blow bubbles, and much of that energy simply sprays out to intergalactic space. But the fearsome glow of radiation and particles from the material accreting around the holes must still impact the surrounding environment. The next step for us is a big one. If we want to complete the story of black holes and the cosmic struggle between construction and destruction, we need to travel even farther. To finish the recipe for apple pie, we need to visit the remotest edges of the universe itself.

6

A DISTANT SIREN

When John Lennon sang that images of broken, dancing light were calling him on and on across the universe, he was thinking metaphorically. But for astronomers and cosmologists, the extremely distant parts of the universe represent a treasure trove of insights into the workings of nature. The finite speed of light is in this case a remarkable gift, opening up billions of years of history for us. As photons pass through the cosmos they carry the imprint of the moment they were emitted, or last reflected. Each one is a messenger that we can query.

If the skies are clear tonight, go outside and take a look around. Perhaps the Moon is visible. But what you see is not *the* Moon, but the past Moon. It's the Moon as it was 1.3 seconds ago. Up there in the sky is another object, tiny but brilliant. It's the planet Jupiter. Or rather, it is Jupiter as it was forty minutes ago. Look around a little more and find the bright stars. If you can see the southern sky, one of the brightest stars of all is Alpha Centauri A, as it was a little over four years ago. Other bright stars are glowing the way they were several decades ago. If the night sky is dark enough you can make out the hazy splash of the plane of the Milky Way galaxy, the projected light of the nearest spiral arms. Most of the light you see has been traveling toward you for thousands of years.

Because of photons' limited speed, we're forever trapped by time, blanketed and shielded from whatever is happening in the cosmos *right now*. But the truth is that it is our minds that have a problem. We need to let go of our conceit that we really witness anything "as it happens." When I drop a coin, I see it hit the ground a few nanoseconds after the coin "thinks" it does. If I watch a seagull scooping up a fish from the distant ocean, the gap in time between its hungry gulp and when I witness it can be tens of nanoseconds. It takes that long for the light reflected from these objects to reach me. The simple fact is that *when* events occur is all relative, something that is of course deeply embedded in Einstein's descriptions of the physical universe. Luckily this characteristic of nature also provides us with the means to practice cosmic paleontology.

In 1962, when astronomer Maarten Schmidt discovered how to interpret the light coming from a distant quasar, he realized that those photons had been en route for 2 billion years. Since that extraordinary measurement, astronomers have striven to push further and further back in cosmic time. We've looked for supernovae, quasars, radio-emitting galaxies, ordinary galaxies, and clusters of galaxies at ever-increasing distances. At times the quest has been highly competitive. Scientists highlight the cosmic distance of a new discovery front and center in the title announcing their work, clearly claiming bragging rights. The following week another team of researchers will try to one-up that discovery by perhaps 100 million years' worth of intergalactic time. Astronomers take pride in their ability to eke out new objects that are fainter and more difficult to characterize than anything that came before. Indeed, it's the nature of the subject, an indelible part of its history and practice.

The result of all these efforts is a cosmic time line. Just like fossil hunters carefully brushing away the dusty grains of rock entombing a specimen, astronomers peel away layer upon layer of spacetime strata. In doing so, we can track the ways in which the populations of stars and galaxies have evolved as the universe has aged. It is how we know that quasars, the most massive and most actively feeding black holes, used to be far more prevalent. It is also how we know that the populations of stars and galaxies of recent times have changed from their earliest days. But why all this galactic evolution happens is still a major question. Theoretical astrophysicists use sophisticated computer simulations in their efforts to understand the steps involved. These virtual worlds incorporate the effects of gravity and gas pressure, and even attempt to model the nature of star formation and the complex interplay of energy and matter in different environments. Yet again and again many of these simulations have produced overgrown galaxies, systems the likes of which are nowhere to be seen in the universe. They contain too many stars, far more than really exist. This is a huge problem, and tells us that we are missing something, some piece of astrophysics that is not yet properly contained in our theories. But we're making progress, and this progress is part of the story here. The bubbles blown by supermassive black holes are one critical component of the answer, but there is still much more at play that we need to understand.

It's in "today's" universe that we've clearly seen the dramatic relationships between supermassive black hole energy output and the nature of galaxies and galaxy clusters. But we need the entire cosmic time line of this phenomenon in order to understand fully how the universe got to be the way it is. Following the fossil clues from the deep past is going to help give us the answers. So let's now take a look at the story of one particular very distant, very strange, and very revealing place in the young universe.

>>>

My entry into the science of supermassive black holes came about because I was interested in something that I thought had nothing to do with these objects at all. For several years I had been pursuing clusters of galaxies across the universe, together with several colleagues around the world. Not literally, of course. Sadly, we were not superheroes. We chased them down the way astronomers chase down any cosmic objects: by surveying the skies with new tools and new persistence. More specifically, we were beachcombers, sifting through the sands of a large database of astronomical X-ray imagery. It all came from the orbiting X-ray telescope called the Roentgen Satellite, or ROSAT for short, a joint European and U.S. space mission. The mission was named after Wilhelm Roentgen, a German scientist distinguished by being the first person to receive the Nobel Prize for physics in its inaugural year of 1901. He was the discoverer of the mysterious "X-rays," or Roentgen rays, that we are now familiar with at our dentist's offices and hospitals. These were produced as a side effect during his laboratory experiments with cathode rays—beams of electrically accelerated electrons. He noticed that despite placing a thin aluminum screen, and then a sheet of cardboard, across the end of his experiment, something was still getting through and producing fluorescence in nearby materials. He called this unknown phenomenon "X-rays." Roentgen was a talented scientist, but little did he know at the time the role that this radiation plays throughout our universe.

During the early 1990s, ROSAT had taken long-exposure digitized images of X-rays emanating from all manner of astrophysical objects—thousands, in fact. As is common practice with almost any large telescope or astronomical instrument, astronomers petitioned the organizations running ROSAT about their favorite things to look at, some known and some exploratory. The telescope would eventually make the observations for the lucky few whose scientific arguments won out in a process of review by their peers. Later on, all this data was placed into a great electronic repository, archived for anyone to use. Scavengers like us could then trawl through the raw material to look for gemstones among the rough.

We were a motley crew. Three of us, myself, the English astronomer Laurence Jones, and the American astronomer Eric Perlman, all ended up spending a couple of years at NASA's Goddard Space Flight Center in suburban Maryland just outside Washington D.C. Jones and Perlman were the first to begin sniffing around in the piles of data from ROSAT, along with Gary Wegner, an astronomer at Dartmouth College. I joined in through my related work on the large structures in the universe, what would later be known as the cosmic web. And, although he was thousands of miles away in Hawaii, we also teamed up with Harald Ebeling, a German-born astronomer with a knack for clever computer algorithms, to sieve through X-ray telescope data looking for interesting objects. Equipped with computers and physics in place of hard hats and lanterns, we were cosmic data miners.

Our goal was to try to find new and ever-more-distant examples of clusters of galaxies. Our clues came from the hot and tenuous gas harbored by the gravity wells of these huge structures. This was the same gas that would later be seen to contain the bubble-blowing black holes in our nearby universe. We looked for the wide, fuzzy smears of X-ray photons from this hot material that had been captured serendipitously in the ROSAT data archives. Once we found these features, we inspected their locations using earthbound telescopes to detect and count the galaxies that might be there. Clusters are as their name implies: their galaxies group together, making a crowded patch on the sky. Many X-ray smears turned out to be just stars or galaxies superimposed by chance, but others turned out to be the real deal: vast collections of galaxies and dark matter in a great gravitational equilibrium.

As time went by we pushed ever farther out into the cosmos, finding these close-knit galactic communities at greater and greater distances. The ultimate prize we sought was to use these objects as a surrogate set of scales for weighing the whole universe: we literally wanted to measure the mass of the cosmos. It was an ambition that many astronomers had long pursued, and in essence the idea is simple. The more mass the universe contains, the more quickly galaxy clusters should appear to grow. Most of that growth should also take place more recently, within the past few billion years, if the universe is chock-full of matter. This means that in a weighty universe we might expect to find very few, if any, clusters, at great cosmological distances. Conversely, in a universe that contains less matter, our measures of cosmic distance and time are different, and the growth of galaxy clusters appears as a weaker and more prolonged affair. By finding lots and lots of clusters, we aimed to refine the statistics and to narrow down estimates of the total content of normal and dark matter in the universe. In doing so, we would arrive at a comparison between science's most fundamental cosmological models and nature itself.

For all of us it became a bit of an obsession. Ebeling, Jones, Perlman, and I seldom went a day without dealing with some piece of the puzzle. After a time we were also joined by Donald Horner, a hardworking graduate student from the University of Maryland. Together we'd pore over the output of the computer algorithms that sniffed through the mountains of X-ray imagery. We'd argue about what looked real and what didn't, flinging printouts of visible-light pictures of what might be galaxy clusters at each other. Then periodically we'd go off to big telescopes to finally nail down our very best candidates, sitting bleary-eyed through long mountain nights in Arizona, Chile, and Hawaii. It wasn't a process that wrapped up quickly, especially since pushing our observations to find ever more distant systems was paramount. There is enormous leverage in distinguishing between different cosmological models if you can see how rapidly clusters form in the young universe. Finding even a single cluster at earlier and earlier cosmic times can make or break certain theoretical scenarios. It's like finding the fossil remains of a feathered dinosaur millions of years before you expect to. Eventually these rare and unique cases might force our ideas about evolution to change.

But galaxy clusters have some tricky characteristics. While I might describe them as "objects" in space, the truth is that they're not like planets or stars. Those bodies are highly self-contained and have a quite well-defined moment at which they finish their formation. A star is a star when its fusion engines are fully started. A planet is a planet when it ceases to accumulate a noticeable amount of material. But a galaxy cluster is a great amoeba of gas and stars, hanging at the intersections of a larger webbing of structure that extends for hundreds of millions of light-years. Matter accumulates almost continually from the surrounding universe, heating as it falls faster and faster inward. Optimistically, we might consider a cluster complete once its constituents reach a state of physical equilibrium. Hot gas will sit quietly in the deepest regions of its gravitational well once the forces of pressure and gravity balance out, and galaxies will remain within the system once their orbits are established. But we know that nature can be awfully messy. The gas cools; black holes throw out energy; incoming matter slowly adds to the system. The precise moment at which a cluster "becomes" itself is therefore open to some interpretation. In astronomy, a field so utterly dependent on observation, nothing beats going and looking. So to answer questions about the baby steps of galaxy clusters, the very best thing would be to find these infants in the act of growing up. And this was where serendipity would rear its head.

Late in the summer of 1999, I made the long journey from the United States to an astronomy conference on the volcanically hewn Greek island of Santorini. It would be tedious to go into how great a place it was for a bunch of sunlight-deprived scientists to hang out, but it was a special thing to be in such dramatic surroundings to discuss the latest science. Also attending was an old colleague, the

English astronomer Ian Smail from Durham University in northern England. Smail had made his name chasing some of the most distant, and hence youthful, objects in the universe. In the latest advance, he and his collaborators were making use of an intriguing and rather new type of astronomical camera. Unlike typical devices for making images of the sky, this camera operated in a netherworld of the electromagnetic spectrum. Between infrared wavelengths and the beginnings of microwaves is a spectral region known as the submillimeter. It's a notoriously tricky regime. At slightly higher energies and shorter wavelengths we can treat light as bouncy photons, trapping and focusing them with our mirrored telescopes. At slightly lower energies and longer wavelengths, we have to treat it as waves that require antennae for detection. Lurking in between is the realm of the submillimeter. It's a slippery beast. (Electromagnetic radiation in this range can penetrate through a person's clothing before reflecting off the outer layers of our skin. Because of this, it is a critical component of some of the machines used to scan your body for concealed items when you go through an airport screening. The stealthy submillimeter photons allow a discreet amount of electromagnetic frisking.)

For astronomy, it's also a regime where our Earth's moist atmosphere is full of potentially confounding noise from the wiggling energetics of water molecules. Submillimeter radiation from cosmic sources is mostly absorbed and overwhelmed by this curtain-like barrier. There are only certain spectral "windows" that are clear enough to look through, and less than half a dozen places on Earth where the environment is dry and dark enough for us to have a hope of peering into the submillimeter universe.

Nonetheless, new technology had produced a camera that could now make astronomical images in this tricky spectral range. The particular device that Smail and his colleagues were using was on the 2.5-mile-high peak of Mauna Kea in Hawaii, attached to the appropriately named James Clerk Maxwell Telescope. It was well known that out in the nearby universe, cold rich dust and gas was an excellent source of submillimeter radiation. The thickly blanketed and mysterious environments of star- and planet-forming nebulae, or dust-rich nearby galaxies, were perfect targets. But Smail and his group were interested in places far, far removed from these.

He and his group wanted to push back into the deep history of the universe, and the submillimeter spectrum offers a unique vantage point. As light traverses the cosmos, its wavelength gets stretched. Our universe is expanding, and spacetime is inexorably swelling. Galaxies far enough apart to have minimal gravitational interaction are flying ever farther away from one another, like raisins in an endless sea of rising yeasty dough. The very tissue of the universe that light must travel through is widening, and a photon that was once at an ultraviolet wavelength can arrive on the other side of the cosmos as one of visible light. A photon that began its journey 10 billion years ago as a blip of infrared energy will arrive as a little wave of submillimeter radiation. Incredibly, Smail's collaborators were finding faint mounds of this cool radiation on the sky that consisted of the cosmically stretched photons of infrared light coming from the rich, dusty, gaseous mix at the very birthplaces of galaxies and stars. These were objects, perhaps protogalactic structures, whose light had taken more than 10 billion years to reach us. As we sat basking in the eight-minute-old photons of brilliant Greek sunshine, we talked excitedly about their most extraordinary finds.

They had taken images of locations that were already known to be strong sources of radio waves in the youthful universe—precisely the kind of objects associated with supermassive black holes in our own more mature cosmic neighborhood. In these places they were discovering great regions of submillimeter emission, tens of thousands of light-years across. These were the hallmarks of clouds of dust that were tens of millions of times the mass of the Sun. They were heated by the intense radiation of newly forming stars, and perhaps something else, shrouded inside. Some of these dusty places were also grouped and clustered together. The statistics used to come to this conclusion were a bit threadbare, but solid enough to take a chance on. It looked like they could be the toddlers that were going to grow into the overfed hulks of galaxy clusters found in our modern universe.

Smail and his colleagues were keen to determine whether or not there really were supermassive black holes lurking in and around their distant sightings of warm material. Such behemoths would also help heat up the masses of dust that were producing the submillimeter light, and would be critical ingredients to understand. I was keen to find out whether these really were the locations of youthful galaxy clusters, the beginnings of those great cathedral-like structures of stars, gas, and dark matter. If they were, then perhaps we could learn something of those structures' initial building blocks.

There was an obvious way to pursue all these goals, and that was through an X-ray telescope. Only the most energetic and penetrating X-ray photons stood a chance of drilling out through the dusty cloak surrounding a massive black hole in such a place. And the fog-like X-ray emission of hot gas getting trapped inside a growing cluster's gravitational bowl was also the best way to measure that great warp in spacetime. We clearly needed to look for X-rays from one of these submillimeter mysteries, and luckily our timing was good. Chandra, NASA's high-performance, $2 billion X-ray space telescope, had just been launched a few months before I met with Smail in Santorini. It was the ideal instrument to chase these distant objects. We just needed to choose the best target. It wasn't a hard choice to make: it had to be the brightest of the distant and dusty blobs, with the uninspiring name of 4C41.17. This mysterious form was a mind-boggling 12 billion light-years away.

It was a long shot that we'd be able to detect anything, but it was a tremendously exciting plan. If we succeeded it would be the most distant detection of this kind of structure yet made, and it really seemed that it was within our grasp. We finally got our chance when we were given the go-ahead after two years of nail-biting anticipation. In late September of 2002, Chandra settled into position to point at our unfathomably distant target. For a total of 150,000 seconds, or about forty hours, it captured and counted X-ray photons streaming in from the universe. Most of these photons were the equivalent of noise on a radio or the fuzzy speckles in an image of a badly tuned TV. X-ray photons, like any other electromagnetic radiation, can traverse the universe. They come from all over; from stars, neutron stars, black holes, great shock waves of supernova explosions, and the hot gas of galaxy clusters. It's a cosmic forest full of rustling leaves. But in among this barrage of random bits and pieces was a copse of noble trees. A grand total of about 150 of these photons had traversed 12 billion years of cosmic time in a direct path from our mysterious 4C41.17.

›››

And so here again is the scene where we began, with pixels on a screen. In this case, from a display in the middle of my desk covered in scattered papers and coffee stains in my office in New York, these particular pixels formed an image. They conveyed the message that Chandra, high above the Earth's surface, had indeed obtained our precious cargo of data. For the last couple of days it had stared at the sky as it silently circumnavigated the Earth. The finest mirrors and instruments that humans could produce had pointed toward a small patch of the cosmos, close to the constellation of Auriga—the Charioteer. In this direction the glorious view across the bows of our Milky Way galaxy took us all the way to 4C41.17, in the deep cosmic past.

It was mid-morning on the island of Manhattan, and the sound of traffic echoed up through its canyons of rock and steel. I stared at the picture on my screen and squinted at the noisy spread of pixels.

This was a preliminary view, before the data were properly processed and massaged to remove spurious features. Fast-moving particles like electrons and protons had ripped through Chandra's frame high in orbit, spewing energy into its sensitive digital camera. This was nothing unusual, just an occupational hazard of space-based astronomy. But there was a shape in that mess of pixels, and I could see it clearly: a pinpoint of X-ray light, along with something else. I sent the image to a printer down the hall and trotted after to grab the hot paper as it spilled from the machine. Subdued by the heat-fused ink, the noisy features in the image were dissipated. There in stark relief was the extraordinary light of something unknown, a thick streak of brightness poking out from either side of a spot of intensity. It looked like dragonfly wings attached to a compact little body, an entomological picture from a long-forgotten era.

My delight at finding something crazy and interesting in the data soon turned to puzzlement. At the center of the image was the bright pinpoint of X-ray light, and this was fairly easy to explain. Its spectrum had the fingerprint of intense X-ray emission from around the accretion disk of a supermassive black hole, buried somewhere inside the thick dust that we knew existed in this system. That problem was solved—there was no doubt now about the presence of a monster in the midst. But there was this other mysterious stuff: the thinly spread wings of light. Translating their length on the image into real distances revealed that altogether they spanned over three hundred thousand light-years from end to end. If this was the X-ray light from hot gas, heated as it flopped into a baby galaxy cluster's gravity well, then it would also have a very particular flavor—that of the bremsstrahlung radiation from hot electrons we encountered before. The spectrum of X-ray photons would obey a particular pattern, and there would be lots of lower-energy photons and few higher-energy ones. Instead, as I wrestled with the data, I found a much more egalitarian spread of energy. That was all wrong—this meant it could not be coming from just hot gas. That wasn't the only thing that was puzzling. If I computed the total power of this radiation, it was a hundred times greater than the X-ray emission of a normal galaxy cluster. This was also at odds with the radiation originating from hot gas in a baby system, yet here was a vast cloud of *something* merrily pumping out X-ray photons. I sent a worried message to Smail: things appeared funny, off-kilter, and I didn't understand.

I knew about hot gas in galaxy clusters, but not enough about strange structures emanating from what had to be a massive black hole on the other side of the visible universe. I started poring over articles in astrophysical journals and cautiously bringing up the data with other colleagues. A few ideas bounced around. Then I noticed two papers that helped shed light on the problem, literally. One of them was recent and written by Dan Schwartz, an astronomer at Harvard. The other one was written in 1966, by Jim Felten and Philip Morrison, two physicists then at Cornell. Felten had played a pivotal role early on in recognizing that galaxy clusters could emit X-ray radiation from their hot gas, and Schwartz was an expert on black hole jets and X-ray astronomy. Despite the time span between these two works, they had something critical in common: both papers talked about the astrophysical manifestations of a phenomenon that I dimly remembered from my undergraduate physics classes, which had the rather nondescript name of "inverse Compton scattering." I soon realized how important this was in explaining our mysterious object in the distant universe.

In 1922, the American physicist Arthur Compton had discovered that X-rays could bounce off freely floating electrons and change their energy, or wavelength, in the process. In essence, this phenomenon is simple. Particles like electrons can interact with photons. If an electron is just sitting quietly and a photon comes zooming along and bounces off it, like a pebble off a rock, then some of the photon's energy will get transferred to the electron, moving it slightly. The photon will carry a bit less energy afterward, shifting to a lower frequency, and the electron will gain a little motion. But if the electron already has a lot of energy, then the outcome is different, and it is the photon that stands to gain.

Out in the cosmos, there are phenomena that can accelerate particles like electrons to huge velocities that are significant fractions of the speed of light. The jets of matter squirting from black holes are an excellent example. These particles carry an exceptional amount of energy of motion, or kinetic energy. Relativity also tells us that the apparent mass of these particles will increase with speed, and that just further boosts the energy they represent—like a sponge getting more massive as it soaks up water. In this case, when a photon happens to bounce or scatter off one of these speedy electrons, then the *inverse* effect can occur. The fast and hot electron gives up some of its energy to the photon, which is a process known as "inverse Compton scattering," and the photon emerges a new, more energetic beast.

The cosmos is filled with photons. We saw this in our map of forever. Many of these are cosmic microwave photons left over from the early universe, and they make up a large fraction of the background electromagnetic soup of the universe. This means that if fast-moving electrons are generated by a phenomenon like a black hole jet, there's a pretty good chance that as they zip along they'll encounter the photons in the cosmic soup and give them a boost of energy. Incredibly, if the electrons move fast enough they can boost a photon from the microwave domain all the way up to the X-ray and even gamma-ray domain. That's like taking a cup of coffee and boosting its temperature high enough to drive the steam turbines of a power plant.

Here was a way to generate X-ray photons with just the spectral signature that I was seeing. All I needed were fast-moving electrons and lots of low-energy photons. It already looked like a supermassive black hole lived in the system. This could certainly be spewing out electrons in the form of fast-moving jets. Indeed, 4C41.17 was a potent source of radio emission, just as one would expect from spiraling relativistic electrons as they splashed into the surrounding universe. I knew that Smail had a map of the radio emission. In it there was a central bright point, and there were two possible dumbbell-shaped zones, almost in line with the wings of X-ray light. It was a good bet that fast-moving electrons were filling these regions.

But what about the supply of those low-energy photons? In the present-day universe there are about 410 cosmic microwave background photons in every cubic centimeter of space at any given time. We found these earlier on, seeping out of our hypothetical sack full of universe. The problem I was now facing was that this was nowhere near enough to account for the huge output in energy we were seeing. This simply didn't represent enough photons to be boosted to an appreciable number of X-rays. I twiddled my thumbs and stared out the window, trying to imagine myself in that distant place, and the pieces suddenly started fitting together.

We already knew that the region we were looking at had a great output of lower-energy infrared photons from hot dust. That could certainly contribute to the reservoir being boosted to higher energies. But there was something else that I'd overlooked, and it was a huge contributing factor—indeed, it made all the difference. It hinged on the fact that the universe itself was very different 12 billion years ago. At that epoch it wasn't yet 2 billion years after the Big Bang, and back then the cosmos was a much smaller place. Spacetime was quite literally more compact; there was much less space between everything. The distance between the baby forms of galaxies was certainly growing, but it was less than a quarter of what it is today. This also meant that the cosmic background photons had not yet been stretched as much as they would be over the next 10 billion years. At that stage in the history of the universe they were almost five times more energetic than they are today, their little wavelengths that much smaller. If I combined these factors I found that these cosmic photons had provided an electromagnetic ocean around 4C41.17 that was more than five hundred times richer than it is now. It was quickly clear that this could be the answer! The thick sea of photons would bounce and scatter off the electrons pouring out from around a black hole, get boosted in energy, and would light up the region with X-rays. It was like shining a searchlight into a fog—the volume of the beam itself would glow with scattered light. The only difference was that this was a beacon we could see across the known universe, and it would only get more efficient the further back in cosmic time we went.

This meant something else, too, something wonderful. If this was indeed the birthplace of a galaxy cluster, then we were witnessing a central black hole blowing bubbles just as its descendants do in the present-day universe. Except the bubbles were not dark voids, they were lit up as they boosted photons into the X-ray band. Push the cosmic clock back far enough, I realized, and you could invert the color scheme of bubbles in a structure, and a negative could become a positive. It was a beautiful and elegant manifestation of basic, fundamental physics. I called Smail in the U.K., bursting with excitement. We might not have seen the hot gas of a baby galaxy cluster directly, but we'd found the stuff inside that gas. We'd found the glowing bubbles driven by the massive heart of a black hole.

Now there were some big questions that we had to tackle. We wanted to know exactly what was happening in this extraordinary environment. Our measurements told us there was lots of warm dust. Altogether this dust represented a hundred million times the mass of the Sun and was made up of tiny, microscopic grains. It was spread out across a hundred thousand light-years. Something was heating it up, perhaps scores of young bright stars, perhaps the supermassive black hole as it wolfed down more matter. The hole was squirting out relativistic particles and inflating bubbles within an unseen medium of gas, and this gas was being gathered up in the growing gravity well of the system. But it also had to be getting dense and cool enough to produce all the big and short-lived stars that were in turn making all the dust, and it had to be feeding the black hole. We were missing a crucial piece, something that would tie it all together, something that would show us exactly where the rest of the matter was in this system.

A few weeks went by and serendipity raised its head again. This time it was in the form of a chance meeting with the Dutch-born astronomer Wil van Breugel, from the Lawrence Livermore National Laboratory and the University of California. Van Bruegel's specialty and passion was tracking down the most distant and massive galaxies. He also happened to have access to the two great Keck telescopes that are perched on Mauna Kea in Hawaii. At an altitude of thirteen thousand feet, these enormous hunks of steel and glass could gulp down photons from across the universe. They were the perfect tools for capturing visible light from ancient structures. Mentioning our investigation provoked a strong response in van Breugel, who told us he had something we'd want to see.

Wil van Breugel and his colleagues had used custom-built light filters to sniff out the photons that came from very specific changes in the energy hierarchy of electrons within hydrogen atoms. Like any element, hydrogen has a number of electromagnetic scents, and this was one of the key ones. If you could hold these atoms in your hand, they would glow with a distinct ultraviolet light. Place them across the universe, though, and the expansion of spacetime would stretch that light out to visible wavelengths. To exploit this, Wil van Breugel's special filters were tuned to catch precisely these photons from various cosmic objects whose distance was already known. What he showed us took our breath away.

He and his colleagues had captured our object, 4C41.17. It had taken one of the Keck telescopes, with its more than thirty-foot-diameter mirror, over seven hours of exposure time to produce an image. That alone was quite an achievement, but it was the view that stunned us. There was the hydrogen gas, recovering from some as yet unknown buffeting, cooling off by emitting photons of ultraviolet light. It was a colossal structure, and in its center were bright clumps and specks, each thousands of light-years across. But they were inside an even larger canopy, stretching out across the same space that our X-ray light came from. And that canopy had sweeping shapes and forms, cusps and spurs of light. What looked like a huge band of dust

Figure 15. The incredible structure of warm hydrogen gas seen around the distant object 4C41.17. End to end, it spans over three hundred thousand light-years. Near the center great bands of darker, dust-rich material seem to wrap around like a belt. The gas coincides with the spread of X-ray light seen by the Chandra Observatory. Also a source of radio emissions and intense infrared light (seen as submillimeter radiation), this remote place is a gigantic, busy construction site for a galaxy and its billions of stars.

seemed to obscure part of the hydrogen gas, as if a belt were tightly adjusted around an overflowing midriff. There was an hourglass shape to the whole thing, hundreds of thousands of light-years across. It gave the impression of matter both coming and going, flowing inward, but also being propelled outward. It was a picture of a tempest from 12 billion years ago. Smail and I knew that we now had the other piece of the puzzle. We just had to put it all together.

We needed to form an accurate mental picture of this distant, primordial environment. A few phone calls later, and over in the city of Leiden in the Netherlands, one of van Breugel's collaborators, Michiel Reuland, found himself tasked with making a visual representation of all our data combined. It was like constructing a painting in translucent layers, each representing a different part of the electromagnetic spectrum, and each overlapping with the others. Experimenting again and again with different color schemes, and by morphing the maps into smooth and simplified forms, Reuland finally came up with a portrait.

It wasn't pretty. In fact, with its artificial neon colors and overlapping shapes, it looked like a colossal mess. I nicknamed it "cosmic roadkill." But despite the aesthetic issues, it was wonderfully informative. Here was just about everything we knew about this remote object in one single view. In the core was a thick shroud of obscuring dust, tens of millions of times the mass of our Sun. It betrayed the intense formation of new stars and new elements hidden within. The vast spread of ultraviolet light from hydrogen gas looked like an unfortunate spill of curdling milk, splattered across three hundred thousand light-years. On a similar scale was the fearsome X-ray glow from cooling but still relativistic electrons as they powered up cosmic background photons and the infrared photons from the warm dust. This X-ray light seemed intertwined with the ultraviolet glow of gas. In a line following the main axis, the main thrust of all this radiation and matter were the dumbbell clouds of radio wave emission. Here the very highest-energy electrons were flowing and corkscrewing outward, glowing with the synchrotron radiation that we see in today's systems. They were still quite fresh, just a few million years old. Their origin could only be from a young supermassive black hole at the very core of this chaotic space. And unseen, but hinted at by the very edges of these structures, there had to be an outer cocoon of gas cooling and flowing inward from the cosmic web.

Particularly intriguing to me was the transfer of energy from the jet-driven particles of the black hole to the cosmic photons in order to produce the glowing X-ray forms. Some of this inverse Compton energy had to be absorbed by the very same hydrogen gas that we were seeing all across the structure in van Breugel's images. That energy could help heat and strip electrons away from this material, ionizing the hydrogen gas and slowing down its cooling. This would in turn put the brakes on the condensation of raw material into stars. As the atoms sought to regain those particles, they would glow with ultraviolet radiation when the captured electrons settled back into the atomic energy levels. This was a new mechanism for a black hole to reach out and tweak the cosmic structure. It would only happen in the youthful universe, where spacetime was still compact enough for the soup of photons to efficiently seed this transfer. If this was correct, then we had found a new way for black holes to sculpt and mold the world around them, to disrupt the formation of new stars and structures.

This evidence, together with the incredible mechanical effects of black hole–inflated bubbles, indicates that black holes play a crucial role in the evolution of cosmic structures across time. They regulate the production of new stars in the giant galaxies of today's universe, and have likely done so throughout cosmic time. And in the very distant past, they could limit how big those galaxies could ever hope to grow in the first place, acting like frustrated farmers trying to keep their weeds under control. What we saw in 4C41.17 was the turbulent youth of a giant galaxy in the beginnings of a great cluster, and despite all the young stars pouring out radiation, and the extraordinary shroud of stellar dust from their brief lives, at the core of it all was still a point of infinite density and relentless force. Because of it this galaxy could never grow beyond a certain size. The more matter fed into its core, the more food there was for the black hole, and the more the black hole would pump out disruptive energy. The X-ray glow of cosmic photons that had been boosted by this energy was as luminous as a trillion suns, and could easily be the tipping point for stemming the growth of the galaxy. It could explain the difference between the number of stars we thought the universe should make in these places and the number that it actually does.

After the rush to analyze and then report our findings, there was enough of a lull to meditate on what we had discovered. While our interpretation seems reasonable, there is always room for future investigation. Regardless, the most remarkable aspect for me was my own shifting sense of what a massive galaxy in the young universe might be like. The mental picture that I'd held up to this point was rather static. Surely something like a galaxy would form serenely and majestically from the gentle condensation of material out of the pristine sea of hydrogen, helium, and dark matter? Of course, we knew that there were quasars out there. They were chewing up matter voraciously and spitting out enough energy to change the balance of atoms, ions, and molecules in the young universe. But they were still quite sparse, and half a century after we'd begun to understand their physical nature, they continued to seem a little disconnected from the galaxies that hosted and fed them. Yes, we also knew that there were generations upon generations of new stars in those early eons. A vigorous amount of stellar birth, life, and death was pumping raw elements into the cosmos. These new components were condensing out as the massive clouds of dust that shrouded so much of the first stellar light, hiding it away from us. But despite all this, the great galaxies still felt like noble and vast castles, steady and reliable.

By stark contrast, in the dull name of 4C41.17 we'd looked into a maelstrom. In a billion or two years this would indeed be another giant galaxy, almost certainly an elliptical fuzzball of hundreds of billions of stars. But we were witnessing a stage where it was a pit of seething radiation and particles, much of that driven by the thrashing forces of a growing supermassive black hole. As much as gravity was trying to draw the galaxy together, it was being resisted by this outflow of energy. Clearly, that resistance was ultimately going to be futile, or else the local cosmos of today would be a very different place. This is part of the very same problem facing our theories and simulations of the growth and evolution of cosmic structures, the simplest of which predict far more stars than we actually find in the universe. What we see in 4C41.17 is a direct clue to the way in which nature limits and restrains the growth of the most massive galaxies. The energy generated by the comparatively microbial specks of the black holes in their cores helps hold them in check.

There was one more thing that this extraordinary colossus on the other side of the universe had to reveal to us, and it came from the environment it occupied. When Smail and his colleagues had taken their original deep images in the lurking submillimeter part of the electromagnetic spectrum, they had found something else. Around the distant radio-emitting galaxies like our 4C41.17, there would often be other glowing mounds that were visible in the pictures. In keeping with the grand scheme of a cosmic web of matter in the universe, these young agglomerations of mass tended to cluster together. See one and you were a little more likely to see another nearby, and another near that, and so on.

What this meant was that in any good X-ray image targeting

one of the distant dusty places, there would likely be others. If they harbored supermassive black holes, then these too might show up as bright points of X-ray light as those energetic photons burrowed their way out through the thick, dusty shrouds. Again, Smail and I picked through the data we had for 4C41.17, as well as examining images of two other structures that were also snapshots of young systems, just a couple of billion years old. To our surprise, not only did we find points of intense X-ray emission apparently inside the dusty shrouds of the young galaxies, but in some cases they came in pairs: twin pinpoints of light next to each other. In one, at the hairy edge of uncertainty in among noisy data, there was evidence of even a triple arrangement—three points of X-ray light peeking out. They were tiny clusters of photons on the screen, just edging into statistical believability, but what might elicit a fatigued shrug from other scientists can excite an astronomer to sleepless nights. Luck and hunches play a huge role in the exploration of the universe.

We realized that we were seeing supermassive black holes that might have originally belonged to separate youthful galaxy structures, stellar teenagers adrift in the young universe. These wayward creatures were now merging and coalescing, setting off frenetic waves of star formation and pumping out dust. Through all this, our X-ray images were penetrating down into their hearts and cores. We know that pairs of supermassive black holes are not something we see much of in today's universe. Only some 4 percent of galaxies are thought to harbor multiple giant black holes. This meant that the ones we had found 12 billion years in the past were likely en route to merging with each other, eventually hiding their origins within a single event horizon. Such arrangements had been posited for some time as a way to smooth out and reorganize the orbits of stars in galactic cores to better match what astronomers were seeing in the nearby universe. Two orbiting black holes are a dynamical blender, reshaping the paths of stars around them.

Given a little more time, perhaps a few million years, these distant giants would find one another through their gravitational pulls. Swirling around as an ever faster orbiting pair, they would eventually combine in a crescendo of gravitational radiation, sending ripples in spacetime ringing out across the universe. It looked like this was clear evidence for one route to growing supermassive black holes in the universe: simply have them eat each other. In doing so, they should also leave their fingerprints on the galaxies that host them, disrupting and rearranging the motions of stars around them, leaving an extremely important set of telltale crumbs around the cookie jar.

›››

Our distant colossus and its brethren represent the extreme end of the galactic flora and fauna. They are the giant trees in the richest parts of the cosmic tropical rain forests. In these environments there is now little doubt that supermassive black holes have played a major, likely dominant, role in sculpting the forms that we see. Twelve billion years ago, and even earlier, they served as regulators and law-keepers to stem the flood of new stars as matter cooled and condensed. Since then they have continued to hold matter at bay. The great bubbles inside clusters have mixed together elements and steadied the transformation of raw hot gas into new stars and planets. But this happens in synchronization with the inflow of that matter, an astonishing symphony of feedback and balance. These great systems breathe in and breathe out.

Elsewhere, off in other galaxies, supermassive black holes are also making their presence felt. But in these other copses and clumps of galactic trees and shrubs, the interplay between construction and destruction is more complex. We happen to find ourselves living in a very large spiral galaxy, the Milky Way. It's an interesting terrain, neither a backwater nor one of the universe's greatest cathedral-like structures—the giant elliptical galaxies within clusters. It's natural to wonder what influence black holes have had on this place, and what role they might continue to play. This brings us to the final part of our story, the search for the origins and nature of our own galactic environment and perhaps even of life itself.

7

ORIGINS: PART I

We know that black holes are end states for matter, but not all of them hide away out of sight. Spacetime itself curves to ridiculous extremes to create the ultimate entrapment at these locations. Yet as cosmic gas, dust, and stars approach the event horizon, their dissolution generates vast amounts of energy that pours noisily back out into the universe. In all of nature, black holes are the most efficient engines for converting matter into energy. This energy plays a vital role across cosmic time: it helps control the production of stars, limits the size of the greatest galaxies, and trims rivers of cooling material into mere rivulets.

Supermassive black holes are also closely related to the sizes of the ancient clouds of stars that surround them. This is true whether the singularity contains a million times the mass of the sun or 10 billion times that amount. It is like judging the size of a pot of honey sitting in a garden by the size of the swarm of bees surrounding it. This relationship exists wherever we find clouds of ancient stars buzzing around the centers of galaxies. It is true for the ellipticals as well as for the majestic spirals, in which clouds or bulges of old stars sit at the center of great disklike wheels of slowly rotating matter. But some galaxies lack this central, puffed-up cloud of old stars; our Milky Way is one such system. In situations like these there are still central black holes, and they can be a million to a hundred million times the mass of the Sun, but somehow the processes at play in other galaxies didn't arise in the same way to link them to central stellar swarms.

In all cases, though, an intimate relationship clearly exists between giant black holes and their host galaxies; they have "coevolved." That's extraordinary, given that they are such disparate structures: one is tens of thousands of light-years across, the other a billion times smaller. From place to place, galaxy to galaxy, this coevolution is also quite varied, suggesting that the particulars of history and circumstance must play vital roles. We can see and smell the signs of fundamental mechanisms at work, but we've yet to join all the dots.

As I asserted at the opening of this book, the presence and behavior of black holes in the universe could very well be connected to the origins of life. It's an outrageous-sounding proposition that the extreme and seemingly remote behavior of black holes has anything to do with the capacity of this universe for life. In order for me to make good on my promise to illuminate that connection, we need to take a careful look at the chain of phenomena that we think go into making stars, planets, and living things before coming back to complete our story about black holes. This inevitably leads us to questions about our own particular cosmic circumstances, and I think the answers are quite surprising.

› › ›

A terrified spider scuttles across the wall while a flower unfurls its petals in a vase. Off in the street a dog idly barks at something real or imagined, and deep in the ocean a school of fish darts and swoops around a cloud of frantically paddling krill. Something slimy grows on the underside of a muddy rock while, together with the

100 trillion bacteria in our guts, we sit in our chairs as electrical pulses zip around our brains. This is life.

Here on Earth it is at once a collection of extraordinarily complex and simple phenomena, involving molecular structures and microscopic machines that organize and reorganize matter in a network of self-sustaining processes. The timescale over which these processes operate stretches from nanoseconds to billions of years. Yet for all this multilayered complexity, the fundamental actions are basic. Energy and matter are exchanged with the environment, and the organization of shape and form, at first on very small scales, is offset by an increase in environmental disorder. A single-celled microscopic organism maintains its cell membranes and internal structure at the expense of plucking and inserting material out of and into its surroundings. Quadrillions of these tiny life-forms can change a planet. They alter its atmosphere and modify its surface chemistry. In effect, they geo-engineer it into something new while building their own ordered cells. Eventually, they may even produce a busy multicellular chicken that leaves behind its own merry trail of disorder in the search for food and energy.

Right now we have only one example of life to study: that which exists on a small rocky planet orbiting a modest star in the 14 billionth year of this universe. There is nothing about the nature of life on Earth, however, that suggests it is anything but a fair sample of the mechanisms that could arise anywhere. For example, terrestrial life consists of carbon, hydrogen, oxygen, and nitrogen, plus some other elements. The characteristics of the chemical bonding among these compounds are such that an extraordinary array of complex and energy-efficient molecular structures can form—from amino acids to DNA. There is no obvious example of an alternative chemical set in the cosmos that can do this.

We don't really know the when, how, or why of life's origins, but it's clear that there are some fundamental prerequisites. The first is the elemental mix necessary to produce biologically important molecules. The second is a location, or sequence of locations, for that chemistry to be incubated in and to ultimately occupy. A third requirement emerges as the wheels of life are set in motion, and that is a supply of energy, whether in the form of raw atomic or molecular materials, or thermal energy, or electromagnetic radiation that can drive chemical reactions. In short, the recipe for life calls for ingredients, pots and pans, and a continually hot oven.

It is in this shopping list that the connections between life and its broader cosmic environment come into sharp focus. Earlier, I described how stars build the heavier elements of the universe. They are nuclear pressure cookers, stuffing protons and neutrons together until they can squeeze in no more. While the primordial elements of hydrogen and helium will always be the most abundant cosmically, the next in line are oxygen and carbon. These are generally made deep in the cores of massive stars, although some of these elements are also produced for short periods in the outer shells of aging stars. The very heaviest elements are produced inside the most massive stars, more than eight times the mass of the Sun, and during their violently explosive deaths as supernovae. Very heavy elements are also produced when objects like white dwarfs are tipped over the edge by more material falling onto them. Pushed through Chandrasekhar's magical limit of quantum pressure support, these dense stellar remnants briefly compress and forge additional elements before spewing them out into the universe in a supernova explosion.

Over lots and lots of time, these heavier elements pollute the interstellar and intergalactic gases as great spills of atomic nuclei, diffusing farther and farther into the depths of space. New stars will condense out of this gas where gravity can overcome the resistance of pressure and energy, and the cycle of stellar formation begins again. As we've seen, this can be quite a battle in some locations, and black holes bear a great responsibility for regulating and limiting this process throughout the cosmos.

The precise details of how this matter condenses into new stars and planets are at the forefront of modern scientific inquiry, not least because we are in the midst of searching for other worlds that might harbor life. We now have the technological means to detect and study planets around other stars, as well as the environments of young, protostellar, protoplanetary systems. This so-called "exoplanetary science" is nothing short of a revolution. Since the Epicurean philosophers of ancient Greece and probably well before, we've questioned whether ours is one of many such worlds in the cosmos. Finally, after centuries of trying, we've indeed begun to discover these other solar systems.

Stars form at the center of thick disks of gas and dust that coalesce from nebulae, not unlike the disks of matter that form around some black holes. These fat platters of material, known as protoplanetary disks, can be a thousand times wider in radius than the distance between the Earth and the Sun. Planets condense and grow out of these disks through a variety of possible routes, complicated by the effects of gravitational dynamics. Over a few million years, what was once a beautifully smooth and pristine wheel of matter becomes pocked and lumpy with these coagulating worlds. At the same time as the planets are forming, the disk of material is also being evaporated away. Flooded by radiation from the new and increasingly hot central star and from neighboring stars, it simply boils off. Astronomers can see this happening, and it fundamentally limits the formation of planets. It is much like the way our earthly seasons change, from fertile spring to slow-growing summer and eventually to winter. The stars and the planets that we are left with are in many senses mere fossils of this episode of intense activity. And intense it is. The major planets of a solar system like our own form in about 30 million years, an extremely short time in the grand scheme of things—a mere 0.3 percent of the lifetime of the parent star. We do not yet understand many details of this process, but our observations of alien systems are revealing vital clues that also have something to say about possible life in these places.

One such signpost is the richness of the elements in a protoplanetary disk. In fully formed exoplanetary systems, we see this on vivid display. The heavy-element content of a star is a good indicator of the elemental mix in the original planet-forming material, and astronomers can measure this quantity through the spectrum of a star's light. It goes hand in hand with the likelihood of finding planets. The more heavy elements we detect, the more planets are likely to exist around that star, *and* the more massive they typically are. This makes a lot of sense. Where there are greater quantities of substances like carbon and silicon, there is more raw material for efficiently forming embryonic planetary bodies.

Water also plays a major role within nascent planetary systems. The relatively high abundance of oxygen across the universe, together with plentiful hydrogen, means that water molecules crop up all over the place. In the disk of material around young stars, water plays a key chemical role within the gases and the youthful chunks of condensing material. When water freezes, it also provides a major source of solid matter that helps drive the gravitational agglomeration of protoplanets. Just as the environment in our solar system gets warmer as we get closer to the Sun, so does the environment in the disk of material around a baby star. Conversely, farther away from these warm inner zones, water freezes into a solid and actually helps accelerate the growth of big chunky planet-like lumps. A major fraction of the solid interiors of planets such as Uranus and Neptune are composed of water ices for this very reason.

The nature of forming planets is also influenced by the eventual size of the star itself. Current research suggests that bigger stars form with bigger disks around them, increasing the potential efficiency of making planets. Astronomers are also finding hints that the chemistry that transpires inside a protoplanetary disk is under the thrall of the parent star. These disk environments are great big cauldrons for all manner of atomic and molecular chemistry. Complex carbon molecules are forming, breaking apart, and being transported around in the disk. Our astronomical observations of young stellar systems reveal lots of chemical mayhem, but in among this are clear hints that systems producing smaller stars may have different chemistry taking place than those growing massive stars. The culprit may simply be the electromagnetic radiation streaming off the still-forming star itself. A big baby star makes more electromagnetic noise than a small one. Photons can both destroy fragile molecules and create pathways for other molecules to form. So the chemical makeup of planets may well be related to the size of their stellar parent, among other things.

Earth happens to be located in an orbit about its parent star that allows for a temperate surface environment. Liquid water can flow freely. Cocktails can be drunk on the beach. We don't understand exactly how critical this really is, but the potential for liquid water on a planet seems to be a reasonable signpost for life. Water is both an essential biochemical solvent and a planet-wide contributor to geophysics and climate. Here too, the size and age of a star is a major factor in determining the orbital regimes that can harbor such a planet.

All this means that getting a world that has the right chemical and energetic richness to produce and sustain organisms hinges on many factors. This doesn't prove that worlds like these are necessarily unlikely or very rare—just that they depend on a whole chain of interlinked steps, some of which we've now taken a quick look at. Our next step is to find the connection between these more local phenomena and those on a truly cosmic scale.

The first place to look is up, straight up, to the galaxies. Each one of these great stellar gathering places in today's universe is a result of billions and billions of years of evolution. Dark matter, gas, dust, and stars coalesce, orbit, bump, explode, waft, and circulate in these systems. But as we've seen, galaxies are not all alike, and their global properties can affect the smaller details significantly. For example, the overall elemental mix available in a galaxy today can have a domino effect on the production of stars and planets. Less elemental richness can mean less-efficient cooling of nebular gas, which means fewer stars will form in the galaxy. That elemental ingredients list can also influence the comparative numbers of big stars and small stars. Fewer heavy elements forged in these stars mean fewer planets form around later generations of stars. And then, to add insult to injury, a dearth of these heavy elements directly impacts the raw chemistry that takes place around forming planets. That space chemistry makes a lot of carbon-based, organic molecules. We don't yet fully understand how complex those molecules get at this stage, or how many of them could end up on the surfaces of new planets—especially the small, rocky Earth-like ones. But they may represent a "prebiotic" mixture for life. Instead of life having to wait millions of years for a young world to build complex molecules in some puddle, such worlds could receive a rich starter mix from space. This is admittedly speculative, but not unreasonable.

You can pick any one of those steps as a potentially critical hurdle for a galaxy to be able to generate the kind of environment that we've evolved out of. We can add several other factors into the mix. An environment subjected to blasts of intense cosmic radiation, whether as photons or particles, may be poorly suited for the growth of complex chemistry and molecules. For example, I'd bet that no Earth-like planet exists inside the jets of feeding supermassive black holes. That would surely be a horrible place for delicate biochemistry. Even being on the sidelines of such an intensely disruptive phenomenon might be detrimental to worlds otherwise suited to harboring life. In more general ways, we've discovered also how black holes can mold the universe around them. The key question now is to find out how this affects the chain of events leading to the formation of stars and planets that have the potential to generate, incubate, and sustain life. To tackle that we have to travel back to the very origins of supermassive black holes themselves.

The most distant quasars exist in a very young universe, barely a billion years old. As we've seen, quasars are products of the appetite of the biggest and best-provisioned black holes. Surrounded by accreting matter, they pump out a prodigious amount of energy. But the age of these systems raises a fundamental question. These supermassive black holes must have formed almost contemporaneously with the first generations of stars in the universe. This is a great puzzle, because the way we think black holes form in today's universe is from the catastrophic collapse of massive stellar remains. Once the mass of a spent stellar core or an object like a neutron star exceeds a certain threshold, there is only one way for it to go: down and in. There is no known pressure force that can resist the shrinking of such an object to inside its event horizon. But this produces a baby black hole only a few times the mass of our Sun. Even if it eats matter at the rate required to power something like a quasar, that amounts to only a few Suns' worth of material a year. With a continual food supply, it would still take hundreds of millions of years to reach supermassive scales. So where could those first giant chasms have possibly come from?

Yet again, the devil is in the details. One theory is that the very first generations of stars in the universe are responsible for producing giant holes. Compared to today's stellar objects, some of these firstborn could be unusually massive, hundreds of times the size of the Sun. The pristine hydrogen and helium gas of the young universe cools less efficiently than today's polluted interstellar gas, so a nebular cloud maintains its pressure and doesn't give way as gravity gathers more and more mass together. This can result in the formation of stellar giants. Once nuclear fusion is triggered, these stars burn quickly and produce black holes. By merging with one another and gulping down surrounding gas, they might grow quickly to supermassive sizes. But we don't know for sure; there may not be enough feedstuff around these holes for them to grow so fast.

Alternatively, under the right conditions the growing mass of a young galaxy could conceivably produce a giant black hole directly in its center. This is a possibility that a number of scientists have studied in detail. Matter pours into the swelling gravity well of an infant galaxy. A sufficiently huge blob of gas may form, collapse under its own weight, and simply speed past all the stages that would otherwise turn it into billions of individual stars. The end result is a directly formed, factory-fresh supermassive black hole. This is awfully tricky, though—absolutely pristine hydrogen and helium and perfect conditions would be required to allow such a lot of extraordinarily dense gas to gracefully condense to a single point.

A third, and arguably more plausible, route has to do with the natural messiness of structure formation in the universe. We are pretty confident that the largest galaxies in the cosmos begin as close-knit groupings of smaller component baby galaxies. These fall together within their mutual gravity well, colliding and merging to eventually settle as a giant galaxy. The colossus 12 billion light-years away with which I began this book represents a stage not so long after that kind of agglomeration.

Supercomputer simulations of these primordial environments indicate that the process of collision and merger of these baby galaxies can generate enormous whirlpools of turbulence. It's not unlike when you pull an oar through water. Behind the paddle the water rushes back into the trough, swirling and churning. These turbulent regions should draw in the material from the colliding galaxies, gathering it in a giant and unstable disk within which spiraling waves direct the gas to the center. Here it is concentrated to a level that pushes it right through the barrier of instability, the critical balance point that James Jeans first determined. Gravity takes over, and a weird star forms that's more than ten thousand times the mass of the Sun. In the blink of a cosmic eye the core of this object gives way, and the matter has fallen inside its event horizon to form the giant seed of a supermassive black hole. It all happens so fast that the rest of the galactic gas has no time to disperse or to condense into skittering stars. This gives the new black hole an opportunity to gobble it up and grow very quickly.

We don't yet know with certainty whether any of these three routes operate in the young universe. Youthful galaxies definitely collide and coalesce. Maybe this helps bring more and more fresh matter into the hungry beaks of the nesting baby black holes, allowing them to bulk up. Perhaps, too, the vast turbulent vortices of colliding galactic pieces can generate the overweight clouds of gas that will swiftly collapse into massive holes. Like the hogging bulk of a cuckoo chick sneakily planted in another's home, they might snatch all the food. Something, for sure, is producing supermassive black holes within the first billion years of existence of the universe. Since this is when the first generations of stars are also produced, there should to be ample opportunity for the properties of the black holes to become linked with the properties of those stars. In some cases, perhaps multiple holes, cosmic neighbors, catch each other in their mutual gravitational pull and merge. We've certainly glimpsed more than one giant hole in systems like the distant bubble-blowing maelstrom 4C41.17 that my colleagues and I got to know so well. Such sticky embraces could both boost the growth of the most massive black holes and leave a calling card on the surrounding stellar swarms.

If a central black hole and the cloud or bulge of stars in a galactic center form contemporaneously, then each may imprint its properties on the other. A big cloud of condensing gas, perhaps swirled into a focusing vortex, could produce both a large black hole and a large set of new stars. A smaller amount of material would produce a lesser black hole and fewer stars. Once a concentration of matter collapses within its event horizon, the whole region quickly shuts down. The outflow of energy from this black hole acts to sweep out any leftover gas and prevent much further growth of anything. In effect, it puts a date stamp on the process. This may be reflected in the relationship we see now between black hole mass and the stars of galactic cores. A similar process might also take place during later episodes of black hole growth: when material from intergalactic space falls into a galaxy, it could trigger star formation while firing up the gravity engine at the center. In this way, the growth of the black hole and the formation of stars could be pushed along in tandem.

If we go even further back in cosmic time, there is something else, an effect that might implicate smaller black holes in the conjoined history of stars and galaxies. As I've noted before, about 380,000 years after the Big Bang the universe cooled enough to become transparent in appearance. Until this time it was opaque, as hot hydrogen and helium nuclei and loose electrons whizzed around and a thick soup of photons scattered back and forth between these particles. At this early time, dark matter was smeared out and diffuse, a shadowy component waiting for gravity to take hold. But as the cosmos cooled to a few thousand degrees, the typical energy of the photons fell below an important threshold. They were no longer prone to being absorbed and rerouted in the clouds of electrons and nuclei that were trying to couple with one another. Bona fide atoms could form without interference and the photons could fly freely across the universe, becoming the cosmic background radiation, the remnants of this hot stage. It was a critical moment in the history of the universe.

For a hypothetical observer, though, it marked the beginning of what was possibly the most monumentally dull episode the cosmos has ever gone through. For roughly the next 100 million years, the universe was dark and increasingly chilly. It was like a particularly bad winter in northern Europe. Astronomers refer to this period as the "dark ages" of the cosmos—with good reason, since there was nothing interesting there: no stars, no galaxies, nothing to light it up. Of course, matter was at work slowly gathering itself down into all its self-imposed hollows and valleys in spacetime, but otherwise, all was quiet.

Eventually gravity got its way. The first stars formed, and their radiation poured out into the pristine void. After a hundred million years of solitude, the primordial gas of the universe was buffeted by energetic photons again. Ultraviolet light stripped electrons from their atoms, and now the cosmos became a great Swiss cheese of cold dark gas full of heated, ionized holes surrounding hotly burning stars. This immediately and irrevocably altered the environment for the formation of the next generations of stars. For astronomers this has been and still is a vital subject of investigation, because what happens next is critical in establishing the entire history of stars and galaxies that leads all the way up to the present day.

And this is where, just possibly, the phenomenon of black holes steps in to dramatically subvert and alter the very building blocks of this newly awakened young universe. In 2011, an intriguing study appeared from a group of astrophysicists led by the Uruguayan-born astronomer Felix Mirabel. Their idea is deceptively simple. Increasingly, as scientists attempt to mimic the physical conditions of this teenage cosmos using sophisticated computer simulations, there is evidence that the first stars may have formed not as individuals, but with brothers and sisters. In today's universe most stars are actually part of a pair or a bigger group. It seems that when fertile conditions exist for the formation of stars, nature finds it easier to form them together, often orbiting around each other. It is a good bet that conditions 13 billion years ago would have produced lots of paired-up, binary stars.

However, no two stars are identical, and it is likely that of two massive stars born as a pair, one will live faster than the other. Once it depletes its nuclear fuel, a big star really has only one way to go, and that is to implode and form a black hole a few times the mass of our Sun. In the right configuration, the black hole can then begin to consume its companion. Stellar matter will be torn off and swept into a disk that accretes around and into the hole. In this familiar process, the frictional heating of the disk releases energy as photons, reaching up to the X-ray regime. It is precisely this scenario that powers our own local black hole prototype system of Cygnus X-1, detected in 1964 by rudimentary X-ray telescopes. In Cygnus X-1 a blue supergiant star is feeding matter into a disk around a black hole some ten times the mass of our Sun.

Mirabel and his colleagues realized that if this pairing of stars and holes indeed occurred at the end of the universal dark ages, the cosmic environment could have been radically altered. X-rays are far more penetrating than ultraviolet photons of light, reaching much farther out into the universe before getting ensnared and absorbed. But they have a similar effect on atoms, ripping off electrons and creating electrostatic carnage. In this scenario, the energy from baby black holes eating up their companion stars floods across vast distances. It fundamentally alters the shape and form of structure in the young universe. Instead of a Swiss cheese topology of cold atoms and molecules filled with hot ionized holes, the cosmos would be cooked more uniformly. This both helps and hinders the formation of new stars. The extra heating of gas by these X-rays could slow down the production of the next batches of stellar objects. But in counterintuitive fashion, X-rays that penetrate deep into the densest cores of baby galaxies can actually encourage the rudimentary chemistry of hydrogen gas, which provides a route for new objects to condense.

This is because pure atomic hydrogen has a hard time cooling down. Atoms may bump and scatter against one another, but there are few ways for that energy of motion to be transferred into electromagnetic radiation that can fly away. It's a different story for hydrogen molecules, though, where two hydrogen atoms are joined together. This molecule can rotate like a bandleader's little baton. It can also wiggle and vibrate like a spring with weights on both ends. So when hydrogen *molecules* bump and bash into each other, some of that energy of motion is transferred into their rotation and vibration. It can then escape as low-energy infrared photons. This provides a new and unique route for the gas to cool off. The energy of motion, the thermal energy, of the gas gets transferred into the wiggling molecules, which in turn spit it out as photons that carry the energy away. For this reason, molecular hydrogen cools much faster than simple, single atoms of hydrogen.

But making hydrogen molecules in situ is a horribly inefficient process. Remarkably, the disruptive influence of X-ray photons is incredibly beneficial to this simple chemistry. X-rays can strip electrons from atoms, and in doing so they provide a jump start for the atomic nuclei to bind together, like an electrostatic lighter fluid. The speedily cooling hydrogen molecules make it far easier for gravity to pull material together, since the gas pressure is reduced. While less-energetic photons can't penetrate into dense clouds of gas, X-rays can. And by making hydrogen into molecules in these dense spots, they put the gas on the fast track to making new stars. This theoretical scenario is certainly plausible. In this case, not only do supermassive black holes play a unique and critical role in sculpting the structures of the universe, but small black holes could also be of fundamental importance at the dawn of stellar astrophysics.

Our conclusion is that the very first black holes—large and small—can leave an imprint on all subsequent stellar generations and galactic environments. The production of new elements and the opportunities for planetary systems all hinge on these early effects, as well as on the long-term behavior of galaxies and the black holes they contain. But not all places are created equal, and most stars that are in the cores of places like galaxy clusters are quite old now. Whatever elements their dying siblings managed to spit out into the void are dispersed in the hot intergalactic gas of these vast gravitational crucibles. Very little of it is recycled back into a state from which it can become new stars or planets. Black holes bear great responsibility for this situation. Ten billion years ago, they restricted and limited what was an explosive growth of stars and elements. Since then, they have continued to hold matter at bay. The multiplicity of black holes that we found in the great dusty mountains of submillimeter-emitting material from 10 billion years ago tallies with their early formation inside the merged splatters of baby galaxies. It also tallies with a picture in which massive black holes often merge with one another, leaving telltale signs in the way that stars are spread across galactic centers. On a much smaller scale, we see aspects of the same behavior in individual galaxies. Those with great swarms of old central stars harbor the most massive black holes. These galaxies are also limited in how many new stars they have been able to make over the past few billion years, and where they could make them.

Some of these places must present a much less fertile terrain, given what we know about the cosmic requirements for life. They may be poorer in condensing elements, and are probably poorer in fresh stars with pristine new worlds. But is this true? Are these locations really unfavorable places for life? The challenge we face is that we only have a single example to serve as context: for now, we know only one planet like Earth. But I would argue that this information still lets us learn something fundamental. We exist in a specific place at a specific cosmic time, in a particular part of a particular galaxy in a particular type of region in the universe. Since that environment is part of the conjoined evolution of black holes and their host galaxies, we can ask what special things link us directly to that history.

8

ORIGINS: PART II

Our existence in this place, this microscopic corner of the cosmos, is fleeting. With utter disregard for our wants and needs, nature plays out its grand acts on scales of space and time that are truly hard to grasp. Perhaps all that we can look to for real solace is our endless capacity to ask questions and seek answers about the place we find ourselves in. That is not such a bad thing. Ignorance is far scarier than knowledge. One of the questions we are now asking is how deeply our specific circumstances are connected to this majestic universal scheme of stars, galaxies, and, of course, black holes. Given that we now see how the origins of black holes and galaxies are intimately linked, and how the subsequent evolution of both is tied together, it's not unreasonable to ask what we might owe to these pinhole punctures in spacetime. Fortunately, we live in an era where we can begin to answer that question.

Our gloriously vast universe contains at least 100 billion galaxies. Generations of careful observation, mapping, and extrapolation have gone into producing this estimate. All but a small handful of these great stellar systems are invisible to the naked human eye. Indeed, the tiniest and dimmest systems are in the majority. Dwarf galaxies, miniature versions of the fuzzy stellar swarms that we call ellipticals, are the most numerous systems in the cosmos. They're incredibly hard to spot, though, since they can consist of as few as a couple of million stars and be only a few hundred light-years across. They're so faint that they vanish out of sight for all but the most persistent and well-equipped observers. The big, more easily seen galaxies are broadly divided. The great disks of spirals are almost always large. They can span hundreds of thousands of light-years, and they can contain a trillion stars. Away from the intense environment of galaxy clusters, they represent more than 70 percent of all large systems. Ellipticals can also be huge, but in number they amount to only about 15 percent of all large galaxies.

We are part of this great intergalactic jungle, and to finish my argument about the relationship of black holes to life in the cosmos, I'm first going to dig deeper into the story of our own very particular watering hole.

› › ›

The Milky Way itself is a big system, even by the standards of spirals. Its 200 billion stars amount to a mass approximately 100 billion times that of our Sun, and its disk stretches across a diameter of 100,000 light-years. Our parent star and our home planet are positioned toward the outer edge of this vast plate, although by no means at the edge of the matter it contains. The visible stars represent just one aspect of a slowly rotating, center-orbiting wheel of dust, gas, and dark matter. Every 210 million years, we complete another circumnavigation of the Milky Way. Since the Sun formed more than 4.5 billion years ago in a long-dissipated clutch of other new stars, we have made this galactic round trip just over twenty times.

Our biggest neighbor is the Andromeda galaxy, separated from the Milky Way by a gaping void of 2.5 million light-years of intergalactic space. Our eyes see only the barest hint of a hazy patch at its location. In truth, its light is spread out across the sky in a great

Figure 16. The spiral galaxy known as NGC 6744, widely considered to be a close match for the structure of our own Milky Way galaxy. It is 30 million light-years from our location.

band some six times the size of the full Moon. It is a giant spiral, but it is quite different from the Milky Way. While our galaxy is still actively producing a few new stars every year, Andromeda has descended into late middle age. It's not without baby stars, but they are forming at one-third to one-fifth the rate that they do in the Milky Way. Andromeda's central cloud of very old stars is also far more prominent than that of the Milky Way. Nestled inside this central stellar hive in Andromeda is a black hole 100 million times the mass of our Sun. Just as in most other galaxies, this hole is one-thousandth the mass of the old stars surrounding it.

In 4 billion to 5 billion years, the curved spacetime containing the masses of Andromeda and the Milky Way will cause them to merge. In fact, they've already started falling toward each other. Although this encounter will happen at a velocity of more than a hundred miles a second, it will not be a collision in the traditional sense of the term. There is so much space between the tiny points of condensed matter in stars that the galaxies will simply drift and flow into each other with little violence. Exactly how intimate this vast embrace will be is unclear, and it will play out over hundreds of millions of years. But eventually the combined content of these two great systems may settle into something resembling an elliptical galaxy, and Andromeda and the Milky Way will be no more.

Regardless of the outcome, by the time this slow collision begins our Sun will have used up the hydrogen fuel in its core, which will contract inward as gravity acts against the diminishing pressure in its center. The shrinking interior will get hotter and will flood the upper layers of the solar atmosphere with radiation, inflating them outward. The Sun will grow to a bloated and gouty red-giant star, engulfing what remains of the inner planets, including Earth. Whether or not our distant descendants are still around to witness these events, they will no doubt mark the end of our birthplace. This tiny scrap of rock and water that took life from microscopic single-celled organisms to beings like us in just a few billion years will be erased. But until then, we have a chance to understand what makes the Milky Way tick, and how it compares to all other galaxies.

› › ›

We live in a time of unprecedented cosmic exploration. The tools of modern astronomy are unlike anything else in the history of our species. We have the technological prowess to construct exquisite instruments like the Chandra X-ray Observatory and the Hubble Space Telescope. We can also co-opt the powers of computer automation and global communications to examine previously unimaginable volumes of the universe. Indeed, much of the astronomy we'll practice in the future will involve a degree of rich mapmaking and information gathering the likes of which we have never seen before. Giant new telescopes will sweep the skies, and every few nights they will produce a record of hundreds of millions of cosmic objects, from stars to galaxies. And they will repeat this again and again. They will do for the universe what something as mundane as a security camera does for a city street: constantly monitoring, constantly growing our library of data, and producing a map of the cosmos of ever-increasing levels of detail in both space and time.

Our early steps toward this new type of astronomy have already begun. One such effort has been the project known as the Sloan Digital Sky Survey. This extraordinary enterprise has surveyed over 35 percent of Earth's night sky since the year 2000, detecting 500 million astrophysical objects. Sloan's modest telescope is just over eight feet in diameter, but it scans across swath after swath of the heavens above New Mexico, with its digital camera recording untold cosmic photons. More than a million of the objects it has captured are galaxies, an incredible sampling of the local universe that penetrates 2 billion years deep into cosmic time. But how to pick through such a wealth of data?

In 2007, a consortium of astronomers launched a project called Galaxy Zoo. The idea was simple, but challenging to implement. The galaxies within the Sloan survey needed to be classified—somehow we needed to assign the correct physical label to all the detected systems in order to extract robust statistical facts about them. The characterizations are familiar: spirals and ellipticals, and subdivisions within these kingdoms. This kind of classification might sound like a straightforward task—surely a computer could be used to "recognize" galaxies. However, nature is tricky, and mistakes made at the level of just a few percent can mess everything up. There is an enormous range of natural variation in structures, as well as confusing quirks—and just plain hiccups in the data. Even very clever computer algorithms can be fooled, especially when you have a million systems to work on.

Since the advent of telescopic astronomy four hundred years ago, the human eye and brain have proved themselves to be remarkable image analyzers. With just a small amount of training and practice, a human being can distinguish galaxy types with incredible efficiency. It's like looking at dried flowers or squashed bugs—after a while you can race through samples with little hesitation. There's a daisy, a lily, a rose, another daisy. There's a beetle, a fly, another fly, a mosquito—the human mind is a brilliant pattern-recognition machine. The real problem facing the Galaxy Zoo project was the sheer scale of its goals. The scientists wanted to classify a *million* galaxies; they also required at least twenty duplicate identifications for each possible galaxy, in order to weed out mistakes. Not even a dedicated group of scientists could find the time or perseverance to accomplish this.

The solution was to harness the power of the human hive, to "crowdsource" astronomy like never before. Soon after it launched, Galaxy Zoo put out a request for volunteers on the Internet. Within a month of its call for human eyes, eighty thousand people had carved out time to look at the million galaxies ten times over. They were scientists, students, bus drivers, retirees, amateur astronomers, kids, athletes, writers, artists, doctors—individuals from all walks of life. It was an amazing example of the joyful and satisfying spirit of cooperation. Just a year later, a staggering 150,000 people had made more than 50 million classifications. The project continues to this day, expanding into the details of galaxies and into new data from the huge archives of the Hubble Space Telescope's two decades in orbit.

Huge sets of data like the information gathered through Galaxy Zoo have allowed astronomers to tackle questions that used to be near-impossible challenges. It is like having census data from an entire continent instead of from a few quirky and obscure neighborhoods. Exactly how we will find ourselves interpreting the results will likely play out over the coming decades, but for now we can ruminate on some of the most compelling discoveries. For us, a key one is the link between galaxy properties and the supermassive black holes that they host. At last we have a way to avoid the confusing peculiarities of individual galaxies and to look instead at how they match up against a million other systems.

We find that there is a significant difference between the supermassive black holes inside elliptical galaxies and those inside spiral galaxies. In today's least-massive elliptical galaxies, the least-massive black holes are also the most active black holes—they are still eating and producing energy. The opposite is true in spiral galaxies: in these systems it is the *most*-massive black holes that are producing the most energy. This sounds like a stunning reversal in behavior until we realize that the most-massive black holes in spiral galaxies are in fact about the same size as the *least*-massive black holes in elliptical galaxies.

We can interpret this to mean the following. In today's universe the most-massive black holes are effectively has-beens. It doesn't matter where they are—most of them have eaten their fill and are certainly not going to light up like quasars again. They are starving. Whatever activity they exhibit is typically modest: enough, for example, to regulate the flow and cooling of matter deep inside a galaxy cluster. The lower-mass black holes, from a few million to a few tens of millions of times the mass of the Sun, are the main players in our surrounding universe. They are still growing, albeit rather gently and sporadically. Thus, the quasars and the great elliptical galaxies have exhausted themselves by leaping out of the starting gate, while the spirals and their more modest black holes have been biding their time. It is the ultimate race between the tortoises and the hares. In fact, some observations suggest that the level of black hole growth we see in these tortoises today is larger than it was a few billion years ago. Only after nearly 14 billion years are they finally hitting their stride.

Where, then, does our galaxy, the Milky Way, sit among these grand tortoises? The answer reveals something quite profound, but first we have to understand how to get there. When astronomers talk about matter being fed into supermassive black holes, they talk about "duty cycles," just like the episodic sloshing of clothes inside a washing machine. The speed of a black hole duty cycle describes how rapidly it changes back and forth from feeding on matter to sitting quietly. The periodic distribution of the great bubbles floating up through clusters of galaxies is an excellent example of a duty cycle made visible. Detecting the presence of black holes is far easier when they are "switched on," and the faster this cyclical behavior, the more black holes you will detect at any instant in a region of the cosmos. It's like being in a completely darkened room full of very hungry mice. If you toss out some pieces of cheese, the fastest runners will quickly scurry from crumb to crumb, and you will count many of them simply by listening. The slow ones take big pauses between snacks, and you will count far fewer at any given moment.

The results of surveys like the Sloan and the Galaxy Zoo indicate that this duty cycle is related to the overall stellar contents of a galaxy. These contents are a critically important clue to the nature of a galactic system. The stars in a galaxy can be reddish, yellowish, or bluish; blue stars are typically the most massive. They are therefore also the shortest-lived, burning through their nuclear fuel in as little as a few million years. This means that if you detect blue stars in the night sky, you're catching sight of youthful stellar systems and the indications of ongoing stellar birth and death. Astronomers find that if you add together all the light coming from a galaxy, the overall color will tend to fall into either a reddish or a bluish category. Red galaxies tend to be ellipticals, and blue galaxies tend to be spirals. In between these two color groups is a place considered to be transitional, where systems are perhaps en route to becoming redder as their young blue stars die off and are no longer replaced. With nary a sense of irony, or indeed color-mixing logic, astronomers call this intermediate zone the "green valley."

Surprisingly, over the past billion years it is the largest green valley spiral galaxies that have the highest black hole duty cycles. They are home to the most regularly growing and squawking giant black holes in the modern universe. These galaxies contain 100 billion times the mass of the Sun in stars, and if you glance at any one of them, you are far more likely to see the signs of an eating black hole than in any other variety of spiral. One in every ten of these galaxies contains a black hole actively consuming matter—in cosmic terms they are switching on and off constantly.

The physical connection between a galaxy being in the green valley and the actions of the central black holes is a puzzle. This is a zone of transition, and most galaxies in the universe are either redder or bluer than this. A system in the valley is in the process of changing; it may even be shutting down its star formation. We know that supermassive black holes can have this effect in other environments, such as galaxy clusters and youthful large galaxies. It might be that these actions are "greening" the galaxies. It might also be that the circumstances causing the transformation of a galaxy are feeding matter to the black hole.

As we study other spiral galaxies in the nearby universe, we do find evidence that the black holes pumping out the most energy have influenced their host systems across thousands of light-years. In some cases, the fierce ultraviolet and X-ray radiation from matter feeding into the holes can propel wind-like regions of heated gas outward. These wash across a galaxy's star-forming regions like hot-weather fronts spreading across a country. Exactly how this impacts the production of stars and elements is unclear, but it's a potent force. Equally, the trigger for such violent output from the central hole can influence the broader sweep of these systems. For example, the inward fall of a dwarf galaxy captured by the gravity well of a larger galaxy stirs up material to funnel it toward the black hole. It is like fanning the embers of a spent fire to relight it. The gravitational and pressure effects of that incoming dwarf galaxy can also dampen or encourage the formation of stars elsewhere in the larger system. Some or all of these phenomena could help link a supermassive black hole to the age (and hence color) of the stars around it.

Remarkably, astronomers have recently realized that our Milky Way itself is one of these very large green valley galaxies. What this means is that our supermassive black hole should be on a fast duty cycle, which is quite a surprise. I've talked about the black hole lurking at the center of our galaxy; it didn't seem so active—in fact, it betrays itself most convincingly by its effect on the orbits of galactic core stars. By this measure, it is only 4 million times the mass of the Sun, a relative whippersnapper. Yet according to our canvassing of the universe, it should also be one of the very busiest.

To paraphrase Humphrey Bogart, of all places in all the galaxies in all the universe, we had to go and find ourselves in this one. It is of course tempting to be skeptical: we haven't thought of our galaxy as playing host to a particularly hungry supermassive black hole. But perhaps this is just a question of timing, of our short lives compared to the lifetime of the cosmos. We need to find out what's going on—do we really live in a quiet or a busy intergalactic neighborhood? Intriguingly, some dramatic evidence now suggests that our received wisdom is due for an overhaul. That evidence comes from viewing the Milky Way through some very special glasses.

› › ›

The most energetic form of electromagnetic radiation is the gamma-ray photon. Gamma rays have wavelengths less than the size of an atom; they are highly penetrating, much more so than X-rays, and will travel through anything but the thickest sheets of metal or rock.

On Earth they originate from the processes occurring within unstable atomic nuclei as part of natural radioactivity. For example, gamma rays produced from the isotope Cobalt-60 are used by the food-processing industry to irradiate and sterilize products like meat and vegetables. Out in the universe, they come from some of the most violent and energy-rich events: stellar implosions, hypersonic shock waves, and the effects of ultra-relativistic particles streaking across space.

For decades a particularly mysterious and persistent set of gamma-ray photons have been finding their way into astrophysical detectors. Although the signal proved hard to pin down, it was clear that these ever-present gammas were coming from a very particular direction—from the inner regions of our own galaxy. It was an ominous sign of fierce processes occurring somewhere deep within the Milky Way.

Eventually, X-ray telescopes, like the Roentgen Satellite we've already encountered, began to pick up tentative signs of immense structures jutting out from our galactic core. These zones of X-ray light were difficult to spot because of their extreme faintness, but astronomers were able to see that they resembled conical funnels opening out toward intergalactic space, spanning thousands of light-years. Their presence suggested that a release of energy, some kind of outflow or vast galactic wind blowing from the inner galactic sanctum, was propelling tenuous hot gas outward.

During the early twenty-first century, astronomers were also charting out the mottled tapestry of the cosmic background radiation through their microwave receivers. These stretched remnants of the photons from the dawn of the universe contained something unusual, too—another tantalizing glimpse of a huge structure. As the scientists analyzed the great microwave sky maps, they saw hints of a subtle tinge, a haze covering that same inner zone of our galaxy. It suggested that the cosmic photons might be passing through some kind of structure composed of fast-moving particles. The photons were being altered on their way to us, their energies shifted by something lurking in this region.

In 2010, a small team from Harvard University led by astronomer Doug Finkbeiner announced a remarkable discovery. Two years earlier NASA had launched a new observatory into orbit. Named Fermi after the famous physicist Enrico Fermi, this instrument represented a huge advance in the way we study gamma rays from space. It could produce high-fidelity gamma-ray images, opening up new cosmic vistas to astronomers. As Fermi orbited the Earth, it constructed a map of the entire sky, scooping up gamma-ray photons from every corner of the universe. Finkbeiner and his team analyzed this map in meticulous detail. They painstakingly combed through it, plucking out all the bright and noisy objects that were blocking the view from our cosmic vantage point. It's like trying to chart the underlying forms of a large and moonlit city. You have to remove the glare of the office windows, car headlights, and streetlights before you can see the outlines of the buildings.

Gradually they peeled away the layers of the chart . . . and there, beneath everything else, they found something quite extraordinary. There was a faint structure in the gamma-ray light coming from the inner galaxy. It was spread across the sky, and it looked exactly like a pair of bubbles. One emerged on either side of the galaxy, to the "north" and to the "south," a vast pair of globe-like wings reaching twenty-five thousand light-years up and away into intergalactic space. Glowing with gamma-ray photons, these bubbles are anchored at their bases to the very core of the Milky Way.

We think that the gamma-ray photons from these structures come from lower-energy photons that are boosted by fast-moving particles such as electrons. This is exactly the mechanism we've seen in the larger structures surrounding the host galaxies of jet-spewing supermassive black holes. It is the process that we found lighting up the colossal bubbles rising from a black hole in the youthful universe. It originates with particles moving close to the speed of light, accelerated from the regions close to an event horizon.

It's still possible that these galactic bubbles are the result of an enormous flurry of stellar birth and death taking place in the galactic core millions of years ago. Such a "burst" of thousands of stars can produce great outflows of radiation and matter that could conceivably produce similar structures. But there is additional evidence indicating that these gamma-ray bubbles really are the signposts of an episode of black hole growth and activity that occurred within the last hundred thousand years.

When we took our journey toward the galactic center, we found a variety of large and intriguing structures, from giant rings of dense gas to other clumps and clouds of material. We've known these to be cold forms, made of frigid molecules sitting in the chill of interstellar space, or tepid and dull clouds of gas. Yet astronomers have found that some of these otherwise dark structures are glowing with X-ray light. This glow has a very particular flavor. It comes from cold atoms of iron that have been agitated until they release X-ray photons. The best explanation is that this agitation is really a form of reflection. X-ray light washes across the cool nebula, where it is absorbed and re-emitted toward us. In this case the gas acts as a giant hazy mirror, and scientists have concluded that the only plausible original source for this reflected radiation is an intensely energy-rich environment at the very core of the galaxy. But because the X-rays we see are echoes off clouds that are three hundred light-years from the galactic center, it means that we are watching a time-delayed playback. From our perspective, something big and powerful in the very core of the galaxy was throwing out a million times more X-ray light three hundred years ago than it is today.

The pieces of evidence are adding up to a compelling picture of our home environment. If the Milky Way obeys the rules that we see in tens of thousands of other galaxies, then it must contain a black hole that is getting fed very regularly. From the gamma-ray bubbles and the ravaged molecular rings of the inner galaxy to the ghostly echoes of X-ray light produced three hundred years ago, there is every reason to believe that we harbor a black hole that is indeed very active. The hole may not be the largest or the most prolific at producing energy when it eats, but it's a busy object, a stormy chasm in our midst. Centuries ago, it burned bright to create the ethereal reflections from the galactic core. Perhaps twenty-five thousand years ago it erupted on an even greater scale to blow the vast bubbles that glow bright in the gamma-ray sky. We should expect the re-ignition of this gravitational engine at any time. If only John Michell or Pierre-Simon Laplace had had a space-borne telescope at their disposal when they looked up to the stars in their scientific quests—the sky in the 1700s would have been rather spectacular!

›››

Clearly, our Milky Way and its central black hole belong to a special club. They hold a distinctive status within today's universe, one that points to a possible connection between the cosmic environment and the phenomenon of life here on Earth. Scientists and philosophers sometimes discuss what are called "anthropic principles." The word *anthropic* is derived from ancient Greek and means that something pertains to humans, or to the period of human existence. Anthropic principles usually tackle the awkward question of whether or not our universe is somehow just right for life to occur. The argument goes that if just a few fundamental physical laws, or physical constants, in the universe were just a bit different, it would have failed to produce life. But we don't currently have good explanations for why the physical parameters of the universe are what they are. So the question stands out: Why did our universe turn out so suitable for life at all? Isn't that incredibly unlikely?

Like many scientists, I grow uncomfortable when faced with these questions. We're determined to try to overcome any prejudice that we are "special" in any way. Just as Copernicus proposed that the Earth is not at the center of the solar system, we are not central to the universe. Indeed, the universe described by Einstein's field equation *has* no meaningful center. But some of the anthropic arguments are trickier to respond to. One possible solution to the discomfort of assigning ourselves a special status hinges on a conceptual and physical picture of nature that allows for multiple realities, or multiple universes. For example, if our universe is merely one of many that exist within a higher-dimensional version of spacetime, then there's no surprise that we exist here. We simply exist in a universe that has the conditions that allow for the phenomenon of life—there is nothing special about it. It's just an island that has the right climate.

That's all quite entertaining stuff, but it also makes us think a little more about exactly what the laundry list of conditions is for life in a universe. It really is striking that the Milky Way, containing us, lands smack-dab in the sweet spot of supermassive black hole activity. It is possible that this is not mere coincidence, and the first question that springs to mind is whether our solar system experiences direct physical ramifications of the activity of a 4-million-solar-mass black hole some twenty-five thousand light-years away. Could it affect the suitability of our suburban galactic neighborhood for life-bearing planets? When our central black hole switches on, eating and pumping out energy, the evidence doesn't suggest that it's enormously bright from our viewpoint. The huge gamma-ray glowing bubbles extending out from the galactic disk definitely indicate some pretty hefty energy production, but not directed toward us. If larger events ever occurred, they must have been in the distant past, perhaps even prior to the formation of our solar system 4.5 billion years ago. Since then, our central monster has likely had only modest physical impact on distant galactic suburbs like those of our solar system.

From the point of view of life, this may be a good thing. A planet like the Earth could be sideswiped by a large increase in ambient interstellar radiation in the form of high-energy photons and fast-moving particles. Radiation can have a deleterious effect on the molecules inside organisms, and it can even affect the structure and chemistry of our atmosphere and oceans. We may be relatively well shielded at 25,000 light-years from the galactic center, but if we lived closer to the galactic core it might be a different story. So the fact that we *don't* live on a planet closer to the core may not be coincidental. Similarly, perhaps we shouldn't be surprised to find ourselves here at this time, rather than billions of years in the past or in the future.

Our galaxy has, like so many others, coevolved with its central supermassive black hole. Indeed, the clues we seek may be less about the question of how our central black hole can directly influence life on Earth, and more about the role it plays as an indicator of the present state of our galaxy in general. The connection between supermassive black holes and their galaxies provides us with a real tool for gauging galactic history. The ferocious quasars of the younger universe are linked to the biggest elliptical galaxies, mostly sitting in the cores of galaxy clusters. These galaxies formed hard and fast and early, the excitable hares in the race. By now their stars are almost all old, and their raw gas is mostly far too hot to form new stars or planets. Other ellipticals, those great dandelion heads of stars, seem to have formed later as galaxies merged. Something along the way has "quenched" their formation of stars. We think that less-violent, but still incredibly powerful, output from supermassive black holes is an excellent candidate for this regulatory role. The spirals with bulges of central stars jutting high above and below the galactic disks also show the signs of an intimate history with their central black holes. They follow some of the same patterns as the ellipticals. In both, the central black hole mass is one-thousandth of the mass of the surrounding stars. Our neighbor Andromeda is one of these systems, its generous stellar bulge covering a black hole more than twenty times the size of ours.

Lower down the pecking order are bulgeless galaxies, like many spirals. Although the Milky Way is a huge galaxy, one of the biggest in the known universe, it harbors a relative pipsqueak of a black hole. The lack of a stellar bulge is a mystery: either the galaxy had less raw material to form from in the first place, or the regulating black hole never really kicked in, or fewer small galaxies and clumps of matter have fallen into the system across time. The incredibly numerous dwarf galaxies also come up short in the black hole department. The true dwarfs of the galactic zoo are quite pitiful things, often with just a few tens of millions of stars or so, and little sign of the gas or dust that will make new ones. Those that are rich in interstellar soup are often so dark, so devoid of stars, that it is as if someone forgot to light the fuse.

Our galaxy still makes stars, at a rate of approximately three solar masses a year. This isn't much on an individual human timescale, but it means that at least 10 million new stars have been born in the Milky Way since our ancestors started walking upright somewhere in the Olduvai Gorge. This is not bad for a place within a universe that is almost 14 billion years old. The giant galaxies of the young universe, blazing with the quasar light from their cores, are in some senses long burnt out. The annoyed belches of their central black holes quench the formation of any new stars; the rippling waves from their flatulent bubbles of relativistic matter prevent material from cooling down and condensing into stellar systems. A tortoise among these hares, the Milky Way still trudges along.

That we live in a large spiral galaxy with very little central stellar bulge and a modest central black hole may be a clue to the type of galaxies best suited to life: ones that did not spend their past building colossal black holes and fighting the demons unleashed in the process. New stars continue to form in a galaxy like ours, but with different vigor from other systems. Most new stars are forming on the edges of the spiral arms as these great circulating waves disturb the disk of gas and dust. They are also forming farther from the galactic center than they used to. Astronomers say that we live in a region of "modest" star formation. Very active star formation produces an awfully messy environment. It builds the massive stars that burn through their nuclear fuel the fastest, ending up as great supernova explosions. Planetary atmospheres can be blasted away or chemically altered by radiation. Fast-moving energetic particles and gamma rays can pummel the surface of a world. Even the flux of ghostly neutrinos released in stellar implosion is intense enough to damage delicate biology. And those are just the moderate effects. Live too close to a supernova and there's a good chance your entire system will be vaporized.

Yet these are also the very mechanisms by which the rich elemental stew inside stars spreads out into the cosmos. This raw material creates new stars as well as planets. They are planets with complex chemical mixtures of hydrocarbons and water, layered and dynamic, stirred by the heat of heavy radioisotopes, with billions of years of geophysics ahead of them. So somewhere in between the zones of forming and exploding young stars and the nursing homes and graveyards of ancient ones is a place that is "just so," and our solar system resides in just such an environment. It is far enough from the galactic center, but not too close to the busy and explosive realms of stars that are forming right now. Of course, all this will change in 5 billion years, when the Andromeda galaxy comes sailing into us.

The connection between the phenomenon of life and the size and activity of supermassive black holes is quite simple. A fertile and temperate galactic zone is far more likely to occur in the type of galaxy that contains a modestly large, regularly nibbling black hole rather than a voracious but long since spent monster. The fact that there are *any* galaxies like the Milky Way in the universe *at this cosmic time* is intimately linked with the opposing processes of gravitational agglomeration of matter and the disruptive energy blasting from matter-swallowing black holes. Too much black hole activity and there would be little new star formation, and the production of heavy elements would cease. Too little black hole activity, and environments might be overly full of young and exploding stars—or too little stirred up to produce anything. Indeed, change the balance at all and you change the whole pathway of star and galaxy formation. As we've seen, even the presence of small black holes, as the universe emerged from its cosmic dark ages, may have helped direct these chains of events.

The entire pathway leading to you and me would be different or even nonexistent without the coevolution of galaxies with supermassive black holes and the extraordinary regulation they perform. The total number of stars in the universe would be different. The number of low- and high-mass stars would be different. The forms of the galaxies would likely be different, and their organization of gas, dust, and elements would almost certainly be different. There would be places that had never been scorched by the intense synchrotron radiation of a supermassive black hole. There would be other places that had never received that jolt, that kick in the pants, that got star or planet formation up and running.

Lots of cosmic phenomena are connected to the existence of life, but some are a little more important than others. Black holes are on that list, and it's because of their unique nature. No other object in the universe is as efficient at converting matter into energy. No other object can act as a great electrical battery capable of expelling ultra-relativistic matter across tens of thousands of light-years. No other object can grow so massive yet still be so comparatively uncomplicated. A black hole is a dent in spacetime described by just three fundamental quantities: mass, spin, and electrical charge.

Take a look at your hand. It contains atoms of carbon, oxygen, and nitrogen that were forged a million miles below the surface of another star billions of years ago. Your hand also contains hydrogen that was present at the very beginning of the universe. All these elements have felt the forces of black holes. And right now, a tiny fraction of the vast electromagnetic sea of photons racing through the universe is reaching down through our atmosphere, hitting those ancient atoms of your flesh. Some of these photons originated in the fearsome spinning of matter around black holes, or from the accelerated jets of particles rushing at near light speed out into the cosmos. We are awash in their radiation, but it is nothing new to the atoms in your hand. As the cosmic dark ages lifted 13 billion years ago, some of the primordial hydrogen in your body was likely buffeted and tickled by the radiation of feeding black holes. Billions of years later, the stars that built your heavy elements existed because of the history of gravity and energy in one zone of what was becoming the Milky Way galaxy.

This fertile corner of the cosmos has been governed by all that has gone on around it, including the behavior of the black hole at our galactic center. The very places that have sealed themselves away from the rest of the universe have served as one of the most influential forces shaping it. We owe so much to them.

9

THERE IS GRANDEUR

> It is interesting to contemplate an entangled bank, clothed with many plants of many kinds, with birds singing on the bushes, with various insects flitting about, and with worms crawling through the damp earth, and to reflect that these elaborately constructed forms, so different from each other, and dependent on each other in so complex a manner, have all been produced by laws acting around us . . . There is grandeur in this view of life, with its several powers, having been originally breathed into a few forms or into one; and that, whilst this planet has gone cycling on according to the fixed law of gravity, from so simple a beginning endless forms most beautiful and most wonderful have been, and are being, evolved.

The English naturalist Charles Darwin wrote these much-repeated words about life on Earth, but they resonate deeply with our twenty-first-century vision of the universe around us. From atoms and molecules forged in successive stellar generations to galaxies of varied shapes, sizes, and hues, layers of cosmic structure have been molded through time by enormously efficient and hungry black holes, gravity's marvelous engines. Human beings arrive in less than an eye blink of all that rich history. Here we are, self-aware organisms contemplating our place in the cosmos, at the improbable crossroads of a multitude of pathways and possibilities. It's easy to see the grandeur of the entangled bank writ large across this universe.

Our modern understanding of nature is by no means complete, but it has become startlingly rich as we probe further into the intricacies and interconnections of the universe. Of all cosmic phenomena, however, it is the most extreme that hold a particular fascination, and black holes are the ultimate extreme. Among all the conceptions of the human mind, we have really outdone ourselves with these objects—they are fantastical, dream-like, and mythological in stature. But they are much more than just a tall tale—they are a vital and active part of all that we see around us.

There is good reason to think that there are hundreds of billions, perhaps trillions, of black holes scattered throughout the universe. They are endpoints for matter, providing critically influential anchors for the environments surrounding them, even though they are minute on a cosmic scale. Imagine that we could sense directly the curvature and distortion of spacetime as if it were a simple three-dimensional landscape. We would find a universe of gently undulating hills, valleys, and small indentations—peppered with the sharpest little pinholes, so fathomlessly deep that the walls plunge out of view as we peer in. It is a most curious topography: elegant curves punctured by fearsome holes that pin down the very fabric of spacetime and flood it with geysers of radiation and particles.

Why does our universe make these dreadful holes in itself? The fundamental laws of physics tell us that spacetime is capable of being constricted and expanded, curved and dragged into motion. These laws also describe the behavior of electromagnetic radiation and the fuzzy and paradoxical quantum world of the subatomic. Together these rules define matter's critical thresholds, the densities and pressures that break through to extreme states. The same principles show us, first, how gravitation builds dense structures out of normal matter. Such gatherings of matter, poised between forces fighting to collapse and expand them, are incubators for chemical and atomic interactions among their contents, and in some cases they get hot enough to ignite nuclear fusion: a star is born. Eventually a few of these objects, stockpiling mass and succumbing to gravity, reach an incredible density that distorts spacetime irreparably. They drop, sink, burrow, and implode all the way out of what we consider to be normal existence. They leave behind fearsome trails in spacetime, like unplugged drains into the underworld, from which even light cannot escape.

The energy that these tormented corners of space spew back into the cosmos affects almost everything that we see. It influences youthful and turbulent galaxies as well as our own quite special Milky Way. Breaking just one of the crisscrossing strands of cosmic history and energy that connect us to black holes could subvert the entire pathway to life here on our small rocky planet. And it is from this little world that we have doggedly pursued an understanding of the universe around us. The story of this pursuit is both inspiring and sobering. Through our struggles with the most rudimentary challenges of survival and our own penchant for conflict, we have nonetheless reached out for greater knowledge. The sobering part is just how little we understand. This is especially true when it comes to the specific and unpredictable real-world consequences of general cosmic rules. The fundamental physical laws give off such an aura of completeness that we may feel that if we know the laws, we know the universe. A Theory of Everything is a popular conceit and certainly a noble aspiration. But any scientist will tell you that the sense of triumph at solving a beautiful equation lasts only as long as you're willing to ignore the colorful wealth of complexity and the endless surprises of combination, permutation, and chance that fill nature.

Black holes are a really good example. Yes, the workings of black holes and the reasons why they occur in this universe are intimately tied to the most fundamental physical laws of relativity, quantum mechanics, and even thermodynamics. But the actual influence of their presence on the basic character of galaxies, stars, and the matter in them is not an obvious consequence. We may imagine the ways in which a black hole can reveal itself, through its gravitational impact on spacetime, its generation of energy, or its influence on surrounding matter, but only discovery can tell us what nature actually does.

There is little doubt in my mind that the influence of energy feedback from black holes has played a key role in shaping the universe into the way we see it today. Even more critically, this phenomenon was a potent force in the early stages of cosmic evolution when the first galaxies were assembling. The present state of our own Milky Way has been influenced by the balances of radiation and particle energy that participated in clearing the pervasive fog of cold hydrogen and helium gas from the universe during its first 100 million years. That process seems intimately linked to the growth of the first supermassive black holes, as does the surprising but undeniable relationship between the size of those objects and the bulging stellar swarms that surround them at the centers of galaxies. Such relationships may also have a deeper connection to the nature of the elusive dark matter that dominates the mass of our universe. But there is something else too, and that is how black holes have contributed to the special, life-friendly circumstances we find ourselves in.

When our own solar system was forming some 4.5 billion years ago out on a peripheral spiral arm of the Milky Way, the environment and elemental richness of our cosmic birthplace would have been very different and perhaps far less fertile were it not for the universal impact of black holes on their surroundings. It's also now apparent that our own galactic center harbors a moderately massive black hole that belongs to a class of systems still gently growing, capturing, and eating matter episodically. Although we are relatively sheltered, a modest wash of radiation out here on the galactic rim every few hundred thousand years could temporarily modify the atmospheric chemistry of a small rocky planet. Even small changes can have big consequences. A little more or less ozone, a little more or less watery precipitation, and the fortunes of a particular organism could take a turn for the better or worse. A few million years down the line, the results could have amplified to dramatically alter the course of evolutionary history on a planet. The stage that the Milky Way is in right now, a seemingly transitional period in the galactic "green valley" before our collision with the Andromeda galaxy in a few billion years, is connected to our central gravity engine. Exactly what this implies is for us to discover.

Astrophysicists have observed and struggled to understand many other fascinating aspects of black holes' impact on their surroundings, such as the nitty-gritty of how matter descends to a black hole, fighting against the outpouring of radiation from material already deep down in the warped spacetime. This game can play out in many ways. A superhot zone called a corona can form above and below a disk of accreting material. Tenuous but scorching matter boils off the disk, filling this region through magnetically controlled channels—an environment much like the surface of our own Sun, but much more extreme. For massive black holes this is a major boost to generating X-ray light, since the great disks on their own are only hot enough to glow in the ultraviolet. We've seen strange pulsations of energy when we've looked deep inside these systems. These rhythmic patterns of outflowing photons betray the ongoing fight between matter and radiation. Like the bubbling skin on boiling milk, radiation pushes matter away from the inner zones around a black hole until it's heated to a disintegrating burst and flops back down, only to begin the cycle again.

Astrophysicists have also tried to understand whether there is a maximum size for black holes. As they grow, they may simply become so good at generating energy that they push away new incoming material, limiting their own size. In effect, a great photon wind blows whenever matter is accreted by the hole, and that wind stops more food from reaching the throat. It's like trying to feed a roaring bonfire. The more fuel you manage to throw on, the farther you need to back away. Such an impasse might occur when a black hole reaches 10 billion times the mass of our Sun, roughly the size of the largest holes yet scrutinized. Pushing yet another limit, some of the most massive black holes appear to spin at close to the maximum rate allowed by physics.

Some data and theories even hint at opportunities for stars to be born within the great gathering disk of material accreting into a hole. Kinks and disturbances in the circulating matter could allow for its localized agglomeration into new objects. Instead of just destroying the arrangement of matter, the black hole environment could conceivably encourage a new start. What a strange and alien environment this might be for the birth of a stellar system. Could there be planets around these stars? We don't know yet, but we can only imagine what the night skies might be like on such worlds. There is also new and intriguing evidence that some black holes, billions of times the mass of the Sun, have been flung out of their parent galaxies. Ejected during the final stages of the merger of black hole pairs, these mysterious objects race outward, escaping their galactic confines for the desolate emptiness of intergalactic space.

Despite the incredible rate of progress in astronomy and our increasing ability to reach out and explore the cosmos, there is one thing we have not yet managed to do: look at a black hole up close and personal. Is it even possible? Clever astronomical techniques have been devised as surrogate probes of the inner recesses of spacetime surrounding the event horizon, but they still offer an incomplete view. One of these methods, owing much to the English astronomer Andrew Fabian, uses the emission of X-ray photons from iron atoms as they swirl about within the accreting material around a black hole. The curved and spinning spacetime imprints a very particular mark on the energy of these photons, which is itself very specific to the iron atoms. Some photons are blueshifted—boosted in energy;

others are reduced in energy and redshifted, but this process is not evenhanded. The extreme speed of the spinning material takes it into a relativistic regime where light from material moving toward us is not only blueshifted to higher energies, but is also enhanced in density. It's called "beaming," or more colloquially, the "headlight effect." The result is that we see far more light coming from the side of a disk spinning toward us than from the side spinning away. Altogether, the photons from around the black hole come out a little tilted and drunken, with energy askew. Add in the other effects of the extremely curved spacetime, and you're left with a dirty fingerprint that can tell us about black hole mass, spin, and the nature of the disk surrounding it. This is a tricky measurement to make, though, and what we see is not an image but a collection of the energies of these battered photons.

Other efforts seek to exploit the innate structure of the cosmos as a medium for receiving messages from the deepest gravitational cusps. If a pair of neutron stars or black holes come to orbit each other closely enough, they can merge in a spectacular crescendo, setting ripples in motion through spacetime itself that transfer energy out into the universe. These are known as gravity waves, and they propagate outward into the cosmos at the speed of light. Gravity waves stretch and squeeze space a minute amount in a distant place like our solar system. As they pass through us they quite literally alter the underlying yardstick of physical dimensions. Physicists and astronomers have been devising systems to try to detect them.

It's an extraordinarily difficult task, but the potential payoff is enormous. Kip Thorne describes it as listening to a gravitational symphony, one that will tell us the masses of the merging black holes and the nature of their merger, and will provide an ultimate test of our mathematical description of the whirling spacetime at their edges. Experiments like the tongue-twisting Laser Interferometer Gravitational-Wave Observatory (thankfully LIGO for short) employ a similar technique to the one used by Michelson and Morley in 1887 to try to detect changes in the time light takes to travel a set path. In this case, however, the path itself is vulnerable to change as gravity waves race through our neighborhood. Unlike that modest-size early experimental device, a LIGO observatory consists of pairs of 2.5-mile-long tubes through which lasers bounce back and forth. Oriented at 90 degrees to each other, they are designed to sense changes in the actual physical distance that the photons have to travel. A passing gravity wave will quite literally alter that length.

Two such observatories exist, working in tandem: one in Hanford, Washington, and the other 1,865 miles away in Livingston, Louisiana. Their level of precision is mind-boggling. Physicists can detect changes in the dimensions of a laser's path that are smaller than one-thousandth of the size of a proton. As they sift through the myriad sources of confusion and noise—even waves crashing on shorelines hundreds of miles away can produce a signal—scientists are getting tantalizingly close to detecting astrophysical events. For example, they are hoping to confirm the prediction that objects such as a pair of small black holes or neutron stars emit a shrieking "chirp" of gravitational energy just as they come together, orbiting furiously around each other. Plans to construct a space-based gravity wave detector employing a similar technique are currently on hold because of budgetary constraints. If it's built, the Laser Interferometer Space Antenna, or LISA, would consist of three spacecraft marking off a triangle in interplanetary space, an astonishing 3 million miles on a side. In astronomers' view, the effort and expense of setting such a remarkable net would be worth it: LISA would be able to hear deep rumbles as supermassive black holes merged in distant galaxies, as well as the hum and buzz of millions of whirling pairs of dense stellar remnants within the Milky Way.

What of the event horizon, though—the very interface between our familiar universe and the one lost within? It is just outside that final gateway that matter gives up the energy that plays such a vital role throughout the cosmos. Observing this directly would be an ultimate victory. We could see exactly how things really work: the accreting disk of material, the twisted coils of a spin-driven jet of particles. Such a close-up would quite literally reveal the inner workings of these gravitational machines. But even the identifiable small black holes in our galaxy are thousands of light-years away—impossibly distant, tiny, and impenetrable. Perhaps. Perhaps not. You need to go talk to the right people.

›››

In the early 1990s, Keith Gendreau spent his time as a graduate student at MIT working on a new type of camera. It wasn't for holiday snapshots, but for capturing X-ray photons from the universe and turning them into an image. It was cutting-edge technology. You took the type of silicon-based device used in digital cameras, a tiny chip divided up into even tinier pixels, and you redesigned and repurposed it. An X-ray photon whacks into the silicon atoms, dumping energy that pushes electrons out of semiconductor stasis. Locate and count those electrons, and you could begin to build an image of the source of the photons. Previous generations of X-ray imaging devices relied on arrays of gas-filled cells and electrostatic grids. Moving to silicon ushered in a whole new era of high fidelity. The only hitch was that you had to do this in space.

Gendreau was helping to construct and calibrate the camera for a joint Japanese-American project called the Advanced Satellite for Cosmology and Astrophysics, or ASCA. Launched into orbit around the Earth in 1993 from the Uchinoura Space Center at the southern tip of Japan, ASCA spent the next eight years gathering images of the X-ray universe before burning up in the atmosphere over the Pacific Ocean. In the meantime, Gendreau moved to NASA's Goddard Space Flight Center in Maryland, just outside Washington, D.C. An irrepressible creator and tinkerer, he was soon helping to lead a new NASA project in its infancy, a mission aiming to do what seemed to be the impossible. Instead of just studying the remote outward effects of black holes in the universe, NASA aimed to directly observe the event horizon itself.

It might seem counterintuitive, as we think that the event horizon is effectively dark nothingness. In fact, it is—except that the space immediately outside the horizon around a feeding black hole is aglow with the final gasps of matter, and the brilliant disk of accreting material can highlight and pinpoint the location of its impending doom. For Gendreau and his fellow scientists, this is the key to seeing a black hole; all you have to do is to make an image of the intense X-ray light flooding from the disk and its immediate surroundings.

The catch is that even for a supermassive black hole, that innermost disk is perhaps only a few light-days across, yet it may be tens of thousands of light-*years* away from us. If you want to look at the event horizon of the 4-million-solar-mass black hole at the center of the Milky Way, you need to be able to see with extraordinary resolution. It's like taking a good image from Earth of a coin on the surface of the Moon, or seeing the individual pixels on a high-definition TV that is, rather inconveniently, more than three thousand miles away.

Building a telescope to do this presents a phenomenal challenge. The physical properties of light itself create an unavoidable hurdle for any kind of astronomical telescope. Light behaves as electric and magnetic waves that are distorted—diffracted—as they pass into optical apertures and lenses. A perfectly clean wave front becomes a messy one, much like water sloshing in through a harbor entrance. This causes an inevitable blurring of the final image. The smaller the diameter of the instrument, the blurrier the image will be. That's why astronomers love to build big telescopes—they'll have a better chance of making a crisp, sharp picture. Understandably, to create an image of a distant event horizon is going to require an *enormous* telescope.

With X-ray photons there is an additional obstacle. There's a reason why we use X-rays to take images of the interior of our bodies: these small-wavelength photons penetrate better than visible-wavelength photons. Without using exotic materials or complex optical tricks, building an X-ray telescope the way you build an optical one is effectively impossible. Instead, astronomers rely on ingenious techniques to bring X-rays together to form an image. One way is to gently coerce the photons into focus, allowing them to skim like skipping stones across the highly polished surfaces of metal-coated silicon. By constructing a series of glassy cylinders within cylinders, like a set of nested Russian dolls, astronomers can coax the X-ray light toward a focal point sensor in a camera.

It works, but it's hard to make these telescopes big enough to form the sharp, high-resolution images we'd like. The clever solution is to make many little telescopes that act like they are all part of one giant telescope. To build an instrument capable of imaging the matter around an event horizon, we can place dozens of smaller telescopes in space, spread out in a great array many tens of miles across. As they gather up X-ray photons, they beam them across the vacuum to a single camera or detector. By merging these photons, and allowing the electromagnetic waves to combine, we can form an image of incredible resolution. Many telescopes become one.

One design for such a system calls for two dozen small X-ray mirrors called "periscopes" to fly in a great swarm a mile wide. Each periscope is a set of perfectly flat surfaces positioned to gently channel the skittish X-ray photons and divert them toward a master "detector" spacecraft. The many-eyed swarm hovers in space, peering into the cosmos. The detector sits twelve thousand miles away and houses a sensitive digital camera that finally captures and measures the X-ray light. It's a megalomaniac telescope, dozens of spacecraft exploded out into a huge flock-like formation deep in interplanetary space.

Talk to Gendreau and other astrophysicists who are just as passionate about pushing the boundaries of technology and knowledge, and you'll come away with the sense that we could really do this. That's despite some enormous technical hurdles. For example, once we launch a spacecraft like this, we need to position it with a precision of a few *ten billionths* of a meter in order for the combined light to come into correct focus. Even the ethereal force of solar radiation, the beating pressure of stellar photons, is enough to disturb such a delicate ballet. The tiny gravitational pull of other solar system planets, like Jupiter, hundreds of millions of miles away can also throw things out of alignment. All that has to be accounted for and corrected for in order for this armada to hold its position in interplanetary space.

Impossible? It's not. Spacecraft engineering and technology for measuring position and orientation have come a very long way since the early days of crude rockets flung skyward. Standing in his laboratory, Gendreau and I sip our steaming-hot coffee and talk of superfluid helium gyroscopes and other tricks that exploit quantum physics to make the necessary measurements to hold a spacecraft steady. Even the engineering required for a ship to reorient itself fractions of a micrometer at a time is under consideration. It's breathtaking that this is within our reach should we choose to place resources into such an enterprise. A mission like this, now given the official and immodest name of Black Hole Imager, or BHI, would let us see through the dust and gas cloaking the core of our own galaxy, or other places like it. It would allow us to peer into the very workings of a black hole.

We could watch as matter spirals inward, observing its textures and behaviors. We might see how a spinning hole launches its great jets of racing particles, how it reaches out into the universe. We would see the precise mechanism of gravity's engines, the origins of the vast outpourings of energy. The BHI would track material as it sweeps around the innermost parts of the accreting disk and as it is caught up in the whirling spacetime itself. It's the ultimate experiment at bath time, peering into the drain as the water and suds vanish with a slurp. The black hole will bend and distort the light we see, a perfect test of our skills in applying the theories physicists first drafted on paper and chalkboard almost a century ago.

Let's suppose that we do it. We build such a remarkable extension of our human senses, and we peer into the pinholes of spacetime that puncture the universe. We will discover surprises that we could never have anticipated. Whatever they are, they will be wonderful. Finally, thousands of generations after our hominid ancestors loped across the plains of Earth, we would be witness to the endpoints of matter in the universe. We have already found matter's starting point. In the mottled haze of cosmic background photons and the faintest recesses of electromagnetic radiation, we see the imprints of the primordial cosmos, the first steps of normal matter into a nearly 14-billion-year history. And in our great particle accelerators we are re-creating the conditions of the universe mere instants after the Big Bang, letting us peer into the exotic fields and particles that are our distant progenitors.

But now, as we stare into the twisted chasms the universe has made in itself, we see the same matter leaving us behind. For all intents and purposes it is sinking into, but also out of, this cosmos. With a final impossibly dim and reddened glimmer, these particles are releasing themselves to eternity as they pass across the event horizon. Yet just before the end of this remarkably long journey, from Big Bang to oblivion within the sheath of an event horizon, matter plays one final role: it gives up what energy it can, and that energy surges back out into the universe to sculpt and color the very environment we occupy.

In a fit of enthusiastic optimism I tell Gendreau that when the BHI sends back its first picture of an event horizon, he and I will take a trip. We'll fly across one of Earth's great oceans and make our way to a small town surrounded by green hills. There we'll go for a walk on the grounds of Thornhill Rectory, looking for a spot where centuries earlier John Michell might have paused to breathe in the fresh air and gaze upward. When we think we've found it, we'll make a little monument, an image in a frame planted in the ground with a spike. Here, at last, the dark stars will have come home.

NOTES

1. DARK STAR

3 *Chandra*: One of NASA's "Great Observatories," ranked as the Hubble Space Telescope's equal in ambition and cost. Launched on July 23, 1999. Information is available at the NASA/Chandra Science Center/Harvard site: http://chandra.harvard.edu.

4 *Twelve billion years*: Travel time for photons from this distant location, corresponding to a cosmological redshift of 3.8 (the ratio between the apparent recession velocity and the speed of light) and a co-moving distance (used in Hubble's law) of about 23 billion light-years for a flat, vacuum energy–dominated cosmological model. In other words: a very long, long way away.

5 *superclusters*: Collections of galaxy clusters and galaxies spanning hundreds of millions of light-years.

5 *Gondwana*: Southern supercontinent believed to have existed from approximately 510 to 200 million years ago, which subsequently broke apart to form Africa, South America, Antarctica, Australia, and India. See, for example, Peter Cattermole, *Building Planet Earth: Five Billion Years of Earth History* (London: Cambridge University Press, 2000).

7 *Bayeux Tapestry*: The sixty-eight-meter-long tapestry narrating the Norman invasion of England in 1066. Halley's comet is depicted, and was likely seen four months prior to the invasion. The seventy-five-to-seventy-

six-year orbital period of the comet was first determined correctly by the English astronomer Edmond Halley in 1705.

10 *past few decades*: Two excellent further sources are Kip Thorne's *Black Holes and Time Warps: Einstein's Outrageous Legacy* (New York: W. W. Norton & Company, 1994), and the book by Mitchell Begelman and Martin Rees, *Gravity's Fatal Attraction: Black Holes in the Universe* (Cambridge: Cambridge University Press, 2nd ed., 2010).

11 *Thornhill*: The Church of St. Michael and All Angels, Thornhill Parish, has both a memorial to John Michell in its tower and a more modern plaque to commemorate his accomplishments. The memorial is notable for its effusive descriptions of his tender and affectionate traits.

11 *personal detail*: More material is now coming to light about Michell, but I have drawn on numerous scraps of information from many different (often online) sources to produce a very modestly detailed portrait. Another source is Sir Archibald Geikie's *Memoir of John Michell, M.A., B.D., F.R.S., fellow of Queens' college, Cambridge, 1749, Woodwardian professor of geology in the university 1762* (Nabu Press, 2010, reprint of original written in 1918), also available online at the University of California Libraries Digital Archives.

11 *works on navigation and astronomy*: One of Michell's best-known contributions to physical science was his role in the invention of the torsion balance with Henry Cavendish. This wonderful device allows the gravitational force between two ball-like masses to be measured (an extraordinary accomplishment given the weakness of the forces involved), and hence for the gravitational force law to be calibrated by measuring the gravitational constant. Cavendish and Michell have each occasionally and interchangeably been referred to as "the man who weighed the world," although Michell died in 1793, some four years before Cavendish made the actual measurement of the Earth's density.

14 *title of his paper*: Michell's presentation is published as a "Letter to Henry Cavendish" by John Michell, *Philosophical Transactions of the Royal Society of London* 74 (1784): 35–57.

16 *Maxwell's work*: These four relationships were published in toto by Maxwell in 1865, following two earlier papers that had laid the

groundwork. James Clerk Maxwell, "A Dynamical Theory of the Electromagnetic Field," *Philosophical Transactions of the Royal Society of London* 155 (1865): 459.

17 *Einstein would later write*: This comment was made by Einstein in 1940 in the article "Considerations Concerning the Fundaments of Theoretical Physics" *Science* 91 (1940): 487.

19 *Michelson-Morley experiment*: Often cited (as it is here) as the best failed experiment ever. But of course it didn't really fail, it was simply so well executed that it revealed the truth. Michelson and Morley's original paper is quite excellent: Albert A. Michelson and Edward W. Morley, "On the Relative Motion of the Earth and the Luminiferous Ether," *American Journal of Science* 34 (1887): 333–45.

21 *special theory of relativity in 1905*: Published by Albert Einstein as "Zur Elektrodynamik bewegter Körper," which translates into English as "On the Electrodynamics of Moving Bodies," *Annalen der Physik* 322, no. 10 (1905): 891.

22 *This simple fact*: These have been discussed over the past century in many excellent accounts beyond those of Einstein himself. To my mind, the flexibility of time is still perhaps the most amazing and bewildering aspect. Little wonder that relativity has also been a topic for philosophers to mull over.

22 general *theory of relativity*: I explore this topic in significantly more detail in chapter 3. To note here, however: there is some confusion in popular accounts about when Einstein published this theory. Although he presented the correct theoretical ideas in 1915, they were in a series of papers that included retractions and corrections of his earlier efforts. In 1916 he finally presented his better-known and more complete discussion and review article, "The Foundation of the General Theory of Relativity," *Annalen der Physik* 49 (1916).

29 *In 1927, Heisenberg*: His paper was "Über den anschaulichen Inhalt der quantentheoretischen Kinematik und Mechanik," translated roughly into English as "On the visualizable [or "intuitive"] content of quantum theoretical kinematics and mechanics," *Zeitschrift für Physik* 43 (1927): 172. The word *anschaulichen* apparently defies easy translation into English.

31 *In the 1920s astronomers*: Throughout this period the great English physicist Sir Arthur Eddington played numerous roles. Not least was his

devastating (and ultimately incorrect) critique later on of Chandrasekhar's 1935 presentation on white dwarfs.

33 *in 1935 Chandrasekhar presented*: Key papers were "The Highly Collapsed Configurations of a Stellar Mass," *Monthly Notices of the Royal Astronomical Society* 95 (1935): 207, and "Stellar Configurations with Degenerate Cores," ibid., 226.

34 *extreme states of matter*: Both nuclear bombs and stellar interiors involve environments where nuclear constituents (protons and neutrons) can become dissociated from their usual place within atomic nuclei. For objects like neutron stars, the need to use general relativity to understand their structure presents an additional challenge.

36 *John Wheeler*: John Archibald Wheeler (1911–2008) was one of the great American theoretical physicists who worked on general relativity. He also worked on the Manhattan Project and mentored many extraordinary scientists, including Richard Feynman and Kip Thorne. Generations of students know him as one of the authors of the seminal textbook titled *Gravitation* with Charles Misner and Kip Thorne (San Francisco: W. H. Freeman, 1973)—all 1,200 pages of it.

36 *NASA Goddard Institute for Space Studies*: One of the least-known NASA outposts, home to some of the best planetary and climate science research going on today. Tom's (the restaurant) is still there, and tourists are often lined up taking pictures of it, because its exterior is used in the opening sequence for the TV show *Seinfeld*.

2. A MAP OF FOREVER

40 *oldest recognizable:* I have taken a small liberty here in making this a statement of fact. It seems that not everyone agrees on the celestial interpretation of these carvings and paintings from tens of thousands of years ago (during the Paleolithic). One interesting reference is the discussion by Amelia Sparavigna, "The Pleiades: The Celestial Herd of Ancient Timekeepers" (online in the physics preprint archives at http://arxiv.org/abs/0810.1592).

41 *range of photon wavelengths:* Sources differ a little on the actual sensitivity range of human eyes, but between about 380 and 750 nanometers is typical. Sensitivity is not uniform, but peaks at around 550

nanometers (green light) and is a combination of the different cone and rod receptor sensitivities in the human retina. For comparison, bees have orange to blue sensitivity along with some ultraviolet sensitivity—spanning wavelengths of approximately 300 to 600 nanometers.

42 *Harlow Shapley*: Many sources exist on Shapley's life and long career. A detailed obituary was published in *Nature*: Z. Kopal, "Great Debate," *Nature* 240 (1972): 429. The American Institute of Physics holds the transcript of an interview with Shapley in 1966 (www.aip.org/history/ohilist/4888_1.html). The Franklin Institute holds details of his life and family (www.fi.edu/learn/case—files/shapley/index.html).

43 *"The Sun is . . ."* Shapley quote is taken from the published results of his globular cluster survey, "Globular Clusters and the Structure of the Galactic System," *Publications of the Astronomical Society of the Pacific* 30 (1918): 42.

43 *modern mapping of the cosmos*: Edwin Hubble, also using the Mount Wilson Observatory, found in the early 1920s that many nebulae were actually galaxies in their own right, separate and very distant from ours—something that Harlow Shapley did not initially agree with. By 1929, Hubble had also shown that they are all moving away from one another—the first direct evidence of the expansion of the universe. Edwin Hubble, "A Relation Between Distance and Radial Velocity Among Extra-Galactic Nebulae," *Proceedings of the National Academy of Sciences of the United States of America* 15 (1929): 168.

44 *Lewis Fry Richardson*: A brief biography by Oliver M. Ashford is available online to subscribers to the Oxford Dictionary of National Biography Index (www.oxforddnb.com/view/article/35739).

55 *web-like cosmic brain*: As maps of the positions of galaxies have grown, we have seen more and more of this "web" structure. It is also a defining characteristic of larger computer simulations that attempt to model the gravitational behavior of dark and normal matter and how structures form and grow on cosmic scales. The phrase "cosmic web," coined in 1996 by University of Toronto astrophysicist Richard Bond, has become generally used by astronomers.

61 *total observable universe*: Although I use this term loosely here, it does actually stem from a more rigorous definition. The "observable universe"

is everything close enough to us for its light to have had time to reach us (though it requires specialized instruments to detect much of it). The accelerating expansion of the universe (our current understanding of the matter) will eventually limit how "far" we can see—there will be stars and galaxies that we simply won't ever know about because their light will be stretched or redshifted too much.

64 *between light and dark*: The complete answer to this question is also a resolution of what is known as Olbers's Paradox. In 1823, the German scientist Heinrich Olbers first posed the question: If the universe is infinite (or at least very large), why is the sky not uniformly bright with the accumulated light of stars from every apparent direction? Many solutions have been proposed over the years, from steady-state cosmologies to opaque universes. The basic answer is that the universe is neither infinite in age nor static in terms of dynamics—its expansion diminishes the light from distant parts. But on more local scales, the absolute number of stars in any given region is critical in determining what we see as light and dark.

3. ONE HUNDRED BILLION WAYS TO THE BOTTOM

65 *Hoover Dam*: In addition to the U.S. Department of the Interior information online (www.usbr.gov/lc/hooverdam), see Michael Hiltzik, *Colossus: Hoover Dam and the Making of the American Century* (New York: Free Press, 2010).

66 *Bureau of Reclamation*: Some excellent resources and short essays about the Hoover Dam are available at the bureau's own website, www.usbr.gov/lc/hooverdam.

67 *Hydroelectricity in Norway*: Many hydroelectric plants in Norway are all but invisible. Natural mountain lakes provide the needed reservoirs of elevated water, and often the only signs of a power station are a series of large pipes running down the side of a mountain or high cliff to connect to a turbine building.

67 *produced his work*: See the notes from chapter 1 for references to Einstein's papers on special and general relativity. See also the excellent discussion by Kip Thorne in *Black Holes and Time Warps: Einstein's Outrageous Legacy* (New York: W. W. Norton & Company, 1994).

71 *work of many others*: It is true that Einstein is the individual who ulti-

mately succeeded in this great mental feat, but he certainly didn't do it in complete isolation. For example, the German mathematician David Hilbert also arrived at a formulation of the field equations, as Einstein did, in late 1915. Hilbert gave Einstein full credit, but it does seem that Einstein benefited from the backdrop of others working on the problem.

73 extremely *rigid and stiff*: Why is spacetime like this? It's the same as asking why gravity is such a weak force compared to others like electromagnetism. We don't really know, but theoretical physicists have some ideas. These include the Randall-Sundrum model, in which the universe is really five-dimensional and the weakness of gravity is due to the fact that we only experience a small part or projection of its properties into our dimensions. A good popular account is by Lisa Randall herself: *Warped Passages: Unraveling the Mysteries of the Universe's Hidden Dimensions* (New York: Ecco, 2005).

75 *found a solution*: The Kerr solution to the field equations is a more general version of Schwarzschild's solution, and of course applies to any spherical mass.

76 *Roger Penrose*: Penrose's original paper that includes the extraction of black hole spin energy is "Gravitational Collapse: The Role of General Relativity," *Rivista del Nuovo Cimento*, special issue I (1969): 252.

77 *nice geranium*: The astute reader might note that neither a falling whale nor a falling pot of geraniums is an original invention. Douglas Adams, as so often is the case, got there first.

81 *a rocket almost thirty feet long*: The vehicles used were Aerobees—small suborbital rockets just about twenty-five feet in height on the launch pad and capable of carrying a payload of about seventy kilograms (150 pounds) to an altitude of 250 kilometers (more than 150 miles).

82 *Riccardo Giacconi*: Received the Nobel Prize in Physics in 2002 "for pioneering contributions to astrophysics, which have led to the discovery of cosmic X-ray sources" (www.nobelprize.org/nobel_prizes/physics/laureates/2002/giacconi.html).

83 *fresh look at Cygnus X-1*: The most recent observations (involving the Chandra X-ray Observatory) have found that the black hole in the Cygnus X-1 system is almost fifteen times the mass of the Sun, and is spinning eight hundred times a second. This makes it one of the largest "small" holes known in our galaxy—a possibly atypical object. See,

for example, Jerome Orosz et al., "The Mass of the Black Hole in Cygnus X-1," *Astrophysical Journal* 742, article id 84 (2011).

85 *Yakov Zel'dovich*: Zel'dovich's paper on energy release through accretion is "The Fate of a Star and the Evolution of Gravitational Energy Upon Accretion," *Soviet Physics Doklady* 9 (1964): 195.

85 *Edwin Salpeter*: Salpeter's paper on energy release through accretion is "Accretion of Interstellar Matter by Massive Objects," *Astrophysical Journal* 140 (1964): 796.

85 *Karl Jansky*: Jansky's results were published as "Radio Waves from Outside the Solar System," *Nature* 132 (1933): 66.

87 *Finally, in 1962, a series*: A key observation made use of lunar occultation of a distant radio source (a quasar) to pin down its location well enough for optical telescopes to target it. See C. Hazard et al., "Investigation of the Radio Source 3C 273 by the Method of Lunar Occultations," *Nature* 197 (1963): 1037.

87 *Maarten Schmidt*: The discovery of the distance (redshift) of the quasar 3C 273 was presented by Maarten Schmidt in "3C 273: A Star-like Object with Large Red-Shift," *Nature* 197 (1963): 1040. Schmidt's own recollection of the discovery in an interview contains more details; the transcript is held by the Center for History of Physics of the American Institute of Physics (www.aip.org/history/ohilist/4861.html).

88 *raging debate*: This is a lengthy story in its own right. The greatest challenge for scientists was to try to explain the colossal energy output that was implied by objects like quasars and by the radio-bright structures being detected. Key figures included the English physicists Fred Hoyle (eventually Sir Fred Hoyle) and Geoffrey Burbidge, who realized that gravitational energy was probably behind these objects, although exactly how was not clear at that point.

89 *otherwise unremarkable galaxies*: It was also known by this time that many galaxies have exceptionally bright centers (nuclei) that can be seen in visible light, as well as some curious spectral characteristics (for example, the so-called "Seyfert" galaxies). A generic name for all such phenomena became "Active Galactic Nuclei," or AGN for short. While this is a term that is always used in modern astronomy, it can also be confusing, since it covers a multitude of situations. I have therefore avoided its use, preferring to be more explicit about the feeding states of black holes.

90 *Donald Lynden-Bell*: In the interest of full disclosure, Donald was one of my advisors while I studied for a Ph.D., so my discussion is undoubtedly colored by that experience. However, I am not alone in my admiration: in 2008 Lynden-Bell and Maarten Schmidt were joint winners of the Kavli Prize in Astrophysics for their work on quasars and black holes. The original paper is by Donald Lynden-Bell, "Galactic Nuclei as Collapsed Old Quasars," *Nature* 223 (1969): 690.

4. THE FEEDING HABITS OF NONILLION-POUND GORILLAS

96 *flattened ring of gas*: The central molecular ring is a quite complex structure. In addition to the ring there are curved filamentary "spokes" emanating from the very center, seen in radio waves. These also appear to be in motion.

97 *colossal black hole*: The object (black hole) and environment at the very center of our galaxy is also known as Sagittarius A*, often shortened to Sgr A*.

97 *technological prowess*: The mass of the Milky Way's central black hole has been estimated by Reinhard Genzel and his group at the Max-Planck-Institut für extraterrestrische Physik, Garching, Germany (the Max Planck Institute for Extraterrestrial Physics) and by the group led by Andrea Ghez at the University of California, Los Angeles. Both have obtained stellar motions at the galactic center that allow for the estimation of the mass and size of the central object. This is a tremendously challenging exercise, given the tiny size of the stellar orbits, from our perspective, and the faintness of the stars at this distance.

100 *history of that effort*: A fascinating but lengthy story. Interestingly, the research of the past fifty or so years into quasars/radio galaxies/"active" galactic centers has tended to be split off into wavelength regimes. Radio astronomers have studied the lobe-like structures and surveyed for bright radio sources across the cosmos. Astronomers who focus on visible light have pursued spectroscopic observations of quasars and galaxies, and so on. Part of the challenge was to somehow tie together the very different apparent behaviors that emanated from the centers of galaxies. Even confirming that an object like a quasar did

indeed sit within a galaxy was difficult, since the quasar light swamped the starlight of the much, much fainter host galaxy. A large part of the answer is that it depends on whether you are looking "edge-on" or straight down toward the central objects. What has become known as the "unified" model or scheme is a physical arrangement thought to be common to most supermassive black holes. The hole itself is surrounded by both a thinner disk of accreting matter (described later in this chapter) and, outside this, a much thicker "donut" or torus of denser gas and dust. Above and below these structures are smaller clumps and clouds of hot gas that can be moving fast. Jets (as you will also see later in this chapter) can emerge from the center. Quasars are seen when the observer is looking almost straight down the central axis—inside the disk and torus.

101 *one-thousandth of the mass*: This relationship is established by measuring the rate at which the central stars in a galaxy are moving around—their statistically typical velocities. Using Newtonian physics, this provides an estimate of the mass of stars in the bulge. A variety of techniques are then used to evaluate the central black hole mass, which is seen to obey the one-thousandth relationship. The astronomical tools and techniques needed to make this measurement really emerged at the start of the twenty-first century. Two key papers are Laura Ferrarese and David Merritt, "A Fundamental Relation Between Supermassive Black Holes and Their Host Galaxies," *Astrophysical Journal* 539 (2000): L9, and Karl Gebhardt et al., "A Relationship Between Nuclear Black Hole Mass and Galaxy Velocity Dispersion," *Astrophysical Journal* 539 (2000): L13.

107 *"static" surface*: The phenomenon whereby matter within this distance of a spinning black hole can appear to be moving around faster than light is known as an extreme version of the Lense-Thirring effect, or frame-dragging, since it is the coordinate frame of spacetime that is being moved around. See, for example, Josef Lense and Hans Thirring, "On the Influence of the Proper Rotation of Central Bodies on the Motions of Planets and Moons According to Einstein's Theory of Gravitation," *Physikalische Zeitschrift* 19 (1918): 156.

108 *Werner Israel*: Werner Israel's discussion of the limiting spin on black holes is "Third Law of Black-hole Dynamics: A Formulation and Proof," *Physical Review Letters* 57 (1986): 397.

113 *Roger Blandford and Roman Znajek*: Blandford and Znajek described this mechanism in their paper "Electromagnetic Extraction of Energy from Kerr Black Holes," *Monthly Notices of the Royal Astronomical Society* 179 (1977): 433.

116 "*synchrotron radiation*": The history of the discovery of synchrotron radiation is described by Herbert C. Pollock, "The Discovery of Synchrotron Radiation," *American Journal of Physics* 51 (1983): 278. Although the discovery was made in 1947, it took astronomers a long time to recognize that the same mechanism was at play in the universe.

119 *Kip Thorne*: Quote from the prologue of his book *Black Holes and Time Warps: Einstein's Outrageous Legacy* (New York: W. W. Norton & Company, 1994, p. 23).

5. BUBBLES

123 *apple pie from scratch*: Quote from *Cosmos*, the television series by Carl Sagan, Ann Druyan, and Steven Soter, and from the book of the same name by Carl Sagan (New York: Random House, 1980).

124 *another fascinating story*: The history of the investigation of the "lumpiness" of the early universe and its measurement through the study of the tiny variations in the cosmic microwave background radiation (together with the evaluation of the distribution of matter in our present-day universe) is indeed a great story. It's one that is still playing out as we probe in ever more detail the physics of the very young universe. An excellent popular account of the first big breakthroughs with the COBE space mission is by John Mather (who later won the Nobel Prize for his work) and John Boslough, *The Very First Light: The True Inside Story of the Scientific Journey Back to the Dawn of the Universe* (New York: Basic Books, revised ed., 2008).

127 *James Jeans*: The work describing the calculations of gravitational instability by James Jeans is "The Stability of a Spherical Nebula," *Philosophical Transactions of the Royal Society of London. Series A, Containing Papers of a Mathematical or Physical Character* 199 (1902).

131 *emerged in the late 1960s*: First discussion by James Felten et al.: "X-rays from the Coma Cluster of Galaxies," *Astrophysical Journal* 146 (1966): 955.

132 *in the mid-1970s*: A key paper discussing cooling gas was based on observations with the Uhuru satellite: Susan M. Lea et al., "Thermal-Bremsstrahlung Interpretation of Cluster X-ray Sources," *Astrophysical Journal* 184 (1973): L105.

133 *"cooling flow"*: The theoretical and observational interpretations of gas cooling in galaxy clusters came together from three main studies: Len Cowie and James Binney, "Radiative Regulation of Gas Flow Within Clusters of Galaxies—A Model for Cluster X-ray Sources," *Astrophysical Journal* 215 (1977): 723; Andrew Fabian and Paul Nulsen, "Subsonic Accretion of Cooling Gas in Clusters of Galaxies," *Monthly Notices of the Royal Astronomical Society* 180 (1977): 479; and William Mathews and Joel Bregman, "Radiative Accretion Flow onto Giant Galaxies in Clusters," *Astrophysical Journal* 224 (1978): 308.

134 *In 1994*: Andrew Fabian, "Cooling Flows in Clusters of Galaxies," *Annual Reviews of Astronomy and Astrophysics* 32 (1994): 277.

135 *10 million degrees*: As data accumulated, the full picture emerged. A good overview is presented by John Peterson et al., "High-Resolution X-ray Spectroscopic Constraints on Cooling-Flow Models for Clusters of Galaxies," *Astrophysical Journal* 590 (2003): 207.

137 *Boehringer used*: This study used the high-resolution imager on the mission known as ROSAT to study Perseus. Hans Boehringer et al., "A ROSAT HRI Study of the Interaction of the X-ray-Emitting Gas and Radio Lobes of NGC 1275," *Monthly Notices of the Royal Astronomical Society* 264 (1993): L25.

138 *a million seconds altogether*: This extraordinary set of data is described by Andrew Fabian et al., "A Very Deep *Chandra* Observation of the Perseus Cluster: Shocks, Ripples and Conduction," *Monthly Notices of the Royal Astronomical Society* 366 (2006): 417. Since then Fabian and his colleagues have obtained even more data that extend their X-ray map outward across the cluster, revealing more structures: Andrew Fabian et al., "A Wide Chandra View of the Core of the Perseus Cluster" (forthcoming in *Monthly Notices of the Royal Astronomical Society*; available as a preprint: http://arxiv.org/abs/1105.5025).

141 *Perseus is not the only*: Work by astronomers such as Brian McNamara has shown many other clusters with bubbles and activity. See, for example, Brian McNamara et al., "The Heating of Gas in a

Galaxy Cluster by X-ray Cavities and Large-scale Shock Fronts," *Nature* 433 (2005): 45.

141 *"flyball" governor*: The method of attachment of this system is sometimes called a conical pendulum, since instead of swinging back and forth, the pendulum mass moves in a circle at the end of its stiff arm.

143 *converted from gas into stars*: Evidence exists that the observed (low) rate of gas cooling implied by X-ray observations is in accord with the number of new stars forming in at least some galaxy cluster cores *if* the star formation efficiency from cooling cluster gas is 14 percent. This would match the universal fraction of normal matter that is in cluster stars. Michael McDonald et al., "Star Formation Efficiency in the Cool Cores of Galaxy Clusters" (forthcoming in *Monthly Notices of the Royal Astronomical Society*; available as a preprint: http://arxiv.org/abs/1104.0665).

6. A DISTANT SIREN

145 *John Lennon*: Paraphrasing lyrics from Lennon/McCartney, "Across the Universe" (from the Beatles' charity album for the World Wildlife Fund, *No One's Gonna Change Our World*, London, Apple Records, 1969).

147 *produced overgrown galaxies*: Also known as the "overcooling problem." See for example A. J. Benson et al., "What Shapes the Luminosity Function of Galaxies?," *Astrophysical Journal* 599 (2003): 38.

148 *ROSAT*: The Roentgen Satellite was developed by Germany, the United States, and the United Kingdom. It was launched in 1990 and turned off in 1999. Like many space-borne instruments, ROSAT had several detectors attached to the end of one main telescope. These included X-ray imaging devices that exploited the electrostatic characteristics of X-ray photons interacting with atoms—either in gases or in solids. In this way, X-ray photons could be converted to electrical signals that could then be used to construct an image.

148 *Wilhelm Roentgen*: In 1895 he discovered that something still emerged after cathode rays (electrons) passed through a thin film of aluminum with a cardboard backing. He noticed that this unknown phenomenon produced fluorescence in material some distance away, and correctly surmised that it represented a new type of ray or radiation.

150 *a process that wrapped up*: The project that became known as the Wide Angle ROSAT Pointed Survey (WARPS for short) started in 1995 and resulted in seven major scientific papers, the most recent in 2009. Along the way we had help from Matt Malkan at UCLA and others. The first paper was Scharf et al., "The Wide-Angle ROSAT Pointed X-ray Survey of Galaxies, Groups, and Clusters. I. Method and First Results," *Astrophysical Journal* 477 (1997): 79.

152 *produced a camera*: The submillimeter camera used in Hawaii was called the Submillimeter Common User Bolometer Array, or SCUBA for short, built by a team at what was then the Royal Observatory in Edinburgh, Scotland.

154 *uninspiring name of 4C41.17*: As with many astronomical objects, this dull name indicates the source of its first detection and its location, 4C being the fourth Cambridge radio survey and 41.17 indicating the angular declination of the object in the Earth's northern sky. The first inklings that this object might represent a baby galaxy cluster were given by Rob Ivison and colleagues: "An Excess of Submillimeter Sources near 4C 41.17: A Candidate Protocluster at Z=3.8?," *Astrophysical Journal* 452 (2000): 27.

157 *Dan Schwartz*: Schwartz's paper was "X-ray Jets as Cosmic Beacons," *Astrophysical Journal Letters* 569 (2002): 23.

157 *Jim Felten and Philip Morrison*: Felten and Morrison's paper was "Omnidirectional Inverse Compton and Synchrotron Radiation from Cosmic Distributions of Fast Electrons and Thermal Photons," *Astrophysical Journal* 146 (1966): 686.

157 *Arthur Compton*: Winner of the Nobel Prize in Physics in 1927. Biography available from the Nobel Foundation: www.nobelprize.org/nobel_prizes/physics/laureates/1927/compton-bio.html.

161 *Wil van Breugel*: The results that Wil showed us came out of his team's more extensive program of observing distant objects. See for example Michiel Reuland et al., "Giant Lyα Nebulae Associated with High-Redshift Radio Galaxies," *Astrophysical Journal* 592 (2003): 755.

165 *report our findings*: We did, and the paper is by Caleb Scharf, Ian Smail, Rob Ivison, Richard Bower, Wil van Breugel, and Michiel Reuland: "Extended X-ray Emission Around 4C41.17 at z=3.8," *Astrophysical Journal* 596 (2003): 105. The "z=3.8" in the title refers to the cosmological redshift (a surrogate for distance) of the light from

this object, which in turn indicates an apparent velocity away from us that is 3.8 times the speed of light. Of course, it is the universe itself that is expanding and stretching the wavelength of the photons to give this impression.

167 *Only some 4 percent of galaxies*: This statistic has been derived from survey data of the local universe. See for example Xin et al., "Active Galactic Nucleus Pairs from the Sloan Digital Sky Survey. I. The Frequency on ~ 5–100 kpc Scales" (in preprint form at http://arxiv.org/abs/1104.0950, 2011). Also, a related work discusses how the gravitational interactions between other galaxies may encourage the feeding of supermassive black holes: Xin et al., "Active Galactic Nucleus Pairs from the Sloan Digital Sky Survey. II. Evidence for Tidally Enhanced Star Formation and Black Hole Accretion" (also as a preprint, http://arxiv.org/abs/1104.0951, 2011).

7. ORIGINS: PART I

171 *helps control the production of stars*: Numerous reviews have now been written in the scientific literature about the relationship of black holes to star and galaxy properties. One useful article is by Andrea Cattaneo et al: "The Role of Black Holes in Galaxy Formation and Evolution," *Nature* 460 (2009): 213.

171 *some galaxies lack*: The nature of the central stellar bulges of galaxies and their black holes is very much at the forefront of current research—and controversy. In particular, why some galaxies, such as our own, have so little central bulge is a bit of a mystery. A good starting point for this discussion is a short summary by Jim Peebles, "How Galaxies Got Their Black Holes," *Nature* 469 (2011): 305.

173 *can change a planet*: This is not an idle comment. It is now clear that the evolution of life (particularly single-celled microbial life, the bacteria and the archaea) is completely intertwined with the surface evolution of the Earth—from chemistry to climate—over the past 4 billion years. An excellent and provocative discussion is by Paul Falkowski, Tom Fenchel, and Edward DeLong, "The Microbial Engines That Drive Earth's Biogeochemical Cycles," *Science* 320 (2008): 1034.

176 *under the thrall*: Although this research on chemistry around stars of different masses is still very new, it is quite compelling, because the

physical explanation makes a lot of sense. For more on this, see Pascucci et al., "The Different Evolution of Gas and Dust in Disks Around Sun-like and Cool Stars," *Astrophysical Journal* 696 (2009): 143.

178 *but not unreasonable*: Indeed it's not. The jury is still very much out on how the surface chemistry developed on the young planet Earth. We do know, however, that the planet was being pelted by a lot of meteoritic material containing a rich mixture of organic and inorganic molecules. Some fraction of the early chemistry must have been due to this extraterrestrial material—the tail end of planet formation itself.

179 *stellar giants*: For the idea that the first stars in the universe were huge, and would give rise to large black hole remains, see, for example, Piero Madau and Martin Rees, "Massive Black Holes as Population III Remnants," *Astrophysical Journal* 551 (2001): L27.

180 *sufficiently huge blob*: Skipping over any true stellar object and going directly to a large black hole: see, for example, Begelman et al., "Formation of Supermassive Black Holes by Direct Collapse in Pregalactic Halos," *Monthly Notices of the Royal Astronomical Society* 370 (2006): 289.

180 *enormous whirlpools*: These simulations and their implications are reported by L. Mayer et al., "Direct Formation of Supermassive Black Holes via Multi-scale Gas Inflows in Galaxy Mergers," *Nature* 466 (2010): 1082. These results and their broader implications are also discussed by Marta Volonteri: "Astrophysics: Making Black Holes from Scratch," *Nature* 466 (2010): 1049.

182 *"dark ages" of the cosmos*: This is a large area of research. One expert is the astronomer Zoltán Haiman of Columbia University, who also gives an excellent overview and discussion of the possible role of smaller black holes in "Cosmology: A Smoother End to the Dark Ages," *Nature* 472 (2011): 47.

183 *Felix Mirabel*: Led the study that is reported by Mirabel et al., "Stellar Black Holes at the Dawn of the Universe," *Astronomy & Astrophysics* 528 (2011).

185 *molecular hydrogen cools much faster*: This is likely critically important in the very young universe. An excellent reference is Zoltán Haiman, Martin Rees, and Abraham Loeb, "H_2 Cooling of Primordial Gas Triggered by UV Irradiation," *Astrophysical Journal* 467 (1996): 522.

8. ORIGINS: PART II

188 *the Andromeda galaxy*: Also often known by its more "official" astronomical name of Messier 31, or M31 for short.

190 *something resembling an elliptical*: Gravitational simulations of the Andromeda/Milky Way collision suggest that this is a possibility. The largest source of uncertainty in what will happen is actually due to our lack of very high-precision measurements of the relative motion of the two galaxies—we can measure Andromeda's velocity toward us very well, but measuring transverse motion is difficult, so we cannot be certain that it is approaching us precisely head-on.

191 *Sloan Digital Sky Survey*: The SDSS finally began in 2000; its genesis was quite prolonged. A primary leader and advocate for the project (which at the time was a very new concept) was the Princeton astronomer Jim Gunn. The SDSS uses a technique known as drift scanning: the telescope remains fixed, and as the Earth rotates, a strip of the sky the width of the instrument's cameras passes by. Data are continually acquired.

191 *project called Galaxy Zoo*: The project has an excellent website (http://zoo1.galaxyzoo.org/), and a great discussion of the history of the project has been written by Forston et al., "Galaxy Zoo: Morphological Classification and Citizen Science" (available as a preprint, http://arxiv.org/abs/1104.5513, 2011).

194 *duty cycle is related*: The specific results are presented by Schawinski et al., "Galaxy Zoo: The Fundamentally Different Co-evolution of Supermassive Black Holes and Their Early- and Late-type Host Galaxies," *Astrophysical Journal* 711 (2010): 284.

196 *astronomers have recently realized*: Several groups of researchers have stated that the Milky Way seems to be a green valley galaxy. It is possible that Andromeda is one as well, albeit a bit more red than green. A nice discussion is by Mutch et al., "The Mid-life Crisis of the Milky Way and M31" (available as a preprint, http://arxiv.org/abs/1105.2564, 2011).

197 *zones of X-ray light*: The center of our galaxy produces all sorts of X-ray emissions, coming from both small and large structures, making it very hard to peel apart the layers. For example, see Snowden et al., "ROSAT Survey Diffuse X-ray Background Maps. Part II.," *Astrophysical Journal* 485 (1997): 125.

198 *In 2010*: The results that revealed the gamma-ray structure in our galaxy are reported by Meng Su, Tracey Slatyer, and Doug Finkbeiner, "Giant Gamma-ray Bubbles from FERMI-LAT: Active Galactic Nucleus Activity or Bipolar Galactic Wind?," *Astrophysical Journal* 724 (2010): 1044.

199 *the X-rays we see are echoes*: There are several lines of evidence for activity from our central black hole, and I have used the X-ray evidence for discussion. One example of this type of reflection observation is given by Ponti et al., "Discovery of a Superluminal Fe K Echo at the Galactic Center: The Glorious Past of Sgr A* Preserved by Molecular Clouds," *Astrophysical Journal* 714 (2010): 732.

9. THERE IS GRANDEUR

211 *maximum size for black holes*: There may indeed be a maximum (excluding the possibility of the merger of two or more already supermassive holes). In our cosmic neighborhood it's around 10 billion times the mass of our Sun, 2,500 times the size of the Milky Way's central black hole. See, for example, Priya Natarajan and Ezequiel Treister, "Is There an Upper Limit to Black Hole Masses?," *Monthly Notices of the Royal Astronomical Society* 393 (2009): 838.

212 *stars to be born*: There is evidence of rings of young blue stars orbiting within three light-years of the central black hole of the Milky Way galaxy, as well as in Andromeda. For example, see Paumard et al., "The Two Young Star Disks in the Central Parsec of the Galaxy: Properties, Dynamics, and Formation," *Astrophysical Journal* 643 (2006): 1011. Theoretical models seem to concur with the possibility of stars forming out of disks around the black holes; see, for example, Bonnell and Rice, "Star Formation Around Supermassive Black Holes," *Science* 321 (2008): 1060.

212 *flung out*: It's really still speculative, but the Chandra X-ray Observatory may have caught just such a thing. See Jonker et al., "A Bright Off-nuclear X-ray Source: A Type IIn Supernova, a Bright ULX or a Recoiling Supermassive Black Hole in CXOJ122518.6+144545," *Monthly Notices of the Royal Astronomical Society* 407 (2010): 645.

213 *known as gravity waves*: These gravitational ripples produce *strain* on spacetime. Waves expected from astrophysical sources (such as merging

black holes) will have polarizations, and will produce a very particular strain pattern that moves free masses back and forth. The frequency of the waves can be quite high, causing perhaps thousands of oscillations a second. The strength of the wave (its amplitude) also drops off with distance from the source.

218 *Black Hole Imager*: At this stage BHI is still a concept, albeit with a number of laboratory test-bed experiments being carried out to investigate the necessary techniques. Two thorough descriptions written by Keith Gendreau and colleagues were submitted to the *United States Astronomy Decadal Review* in 2010: "The Science Enabled by Ultrahigh Angular Resolution X-ray and Gamma-ray Imaging of Black Holes" (http://maxim.gsfc.nasa.gov/documents/Astro2010/Gendreau_BlackHoleImager_CFP_GAN_GCT.pdf), and "Black Hole Imager: What Happens at the Edge of a Black Hole?" (http://maxim.gsfc.nasa.gov/documents/Astro2010/Gendreau_BHI.pdf).

INDEX

Page numbers in *italics* refer to illustrations.

ILLUSTRATION CREDITS

55 Atlas image obtained as part of the Two Micron All Sky Survey (2MASS), a joint project of the University of Massachusetts and the Infrared Processing and Analysis Center / California Institute of Technology, funded by the National Aeronautics and Space Administration and the National Science Foundation

89 Courtesy of National Radio Astronomy Observatory / Associated Universities, Inc. / National Science Foundation, and investigators R. Perley, C. Carilli, and J. Dreher

98 Courtesy of National Radio Astronomy Observatory / Associated Universities, Inc. / National Science Foundation

104 Courtesy of NASA/JPL-Caltech/R. Hurt (SSC)

106 NASA/ESA/STScI and Walter Jaffe (Leiden), Holland Ford (STScI/JHU)

118 NASA and the Hubble Heritage Team (STScI/AURA). Acknowledgment: J. A. Biretta, W. B. Sparks, F. D. Macchetto, E. S. Perlman (STScI)

139 NASA/Chandra X-ray Science Center (CXC)/Institute of Astronomy (IoA)/Andrew Fabian et al.

162 With thanks to Wil van Breugel and Michiel Reuland. Image obtained with the Keck II ten-meter telescope on Mauna Kea in Hawaii.

189 European Southern Observatory [ESO], image taken with the Wide Field Imager on the MPG/ESO 2.2-meter telescope at La Silla.

ALLEN LANE

an imprint of

PENGUIN BOOKS

Recently Published

Hooman Majd, *The Ministry of Guidance Invites You to Not Stay: An American Family in Iran*

Roger Knight, *Britain Against Napoleon: The Organisation of Victory, 1793-1815*

Alan Greenspan, *The Map and the Territory: Risk, Human Nature and the Future of Forecasting*

Daniel Lieberman, *Story of the Human Body: Evolution, Health and Disease*

Malcolm Gladwell, *David and Goliath: Underdogs, Misfits and the Art of Battling Giants*

Paul Collier, *Exodus: Immigration and Multiculturalism in the 21st Century*

John Eliot Gardiner, *Music in the Castle of Heaven: Immigration and Multiculturalism in the 21st Century*

Catherine Merridale, *Red Fortress: The Secret Heart of Russia's History*

Ramachandra Guha, *Gandhi Before India*

Vic Gatrell, *The First Bohemians: Life and Art in London's Golden Age*

Richard Overy, *The Bombing War: Europe 1939-1945*

Charles Townshend, *The Republic: The Fight for Irish Independence, 1918-1923*

Eric Schlosser, *Command and Control*

Sudhir Venkatesh, *Floating City: Hustlers, Strivers, Dealers, Call Girls and Other Lives in Illicit New York*

Sendhil Mullainathan & Eldar Shafir, *Scarcity: Why Having Too Little Means So Much*

John Drury, *Music at Midnight: The Life and Poetry of George Herbert*

Philip Coggan, *The Last Vote: The Threats to Western Democracy*

Richard Barber, *Edward III and the Triumph of England*

Daniel M Davis, *The Compatibility Gene*

John Bradshaw, *Cat Sense: The Feline Enigma Revealed*

Roger Knight, *Britain Against Napoleon: The Organisation of Victory, 1793-1815*

Thurston Clarke, *JFK's Last Hundred Days: An Intimate Portrait of a Great President*

Jean Drèze and Amartya Sen, *An Uncertain Glory: India and its Contradictions*

Rana Mitter, *China's War with Japan, 1937-1945: The Struggle for Survival*

Tom Burns, *Our Necessary Shadow: The Nature and Meaning of Psychiatry*

Sylvain Tesson, *Consolations of the Forest: Alone in a Cabin in the Middle Taiga*

George Monbiot, *Feral: Searching for Enchantment on the Frontiers of Rewilding*

Ken Robinson and Lou Aronica, *Finding Your Element: How to Discover Your Talents and Passions and Transform Your Life*

David Stuckler and Sanjay Basu, *The Body Economic: Why Austerity Kills*

Suzanne Corkin, *Permanent Present Tense: The Man with No Memory, and What He Taught the World*

Daniel C. Dennett, *Intuition Pumps and Other Tools for Thinking*

Adrian Raine, *The Anatomy of Violence: The Biological Roots of Crime*

Eduardo Galeano, *Children of the Days: A Calendar of Human History*

Lee Smolin, *Time Reborn: From the Crisis of Physics to the Future of the Universe*

Michael Pollan, *Cooked: A Natural History of Transformation*

David Graeber, *The Democracy Project: A History, a Crisis, a Movement*

Brendan Simms, *Europe: The Struggle for Supremacy, 1453 to the Present*

Oliver Bullough, *The Last Man in Russia and the Struggle to Save a Dying Nation*

Diarmaid MacCulloch, *Silence: A Christian History*

Evgeny Morozov, *To Save Everything, Click Here: Technology, Solutionism, and the Urge to Fix Problems that Don't Exist*

David Cannadine, *The Undivided Past: History Beyond Our Differences*

Michael Axworthy, *Revolutionary Iran: A History of the Islamic Republic*

Jaron Lanier, *Who Owns the Future?*

John Gray, *The Silence of Animals: On Progress and Other Modern Myths*

Paul Kildea, *Benjamin Britten: A Life in the Twentieth Century*

Jared Diamond, *The World Until Yesterday: What Can We Learn from Traditional Societies?*

Nassim Nicholas Taleb, *Antifragile: How to Live in a World We Don't Understand*

Alan Ryan, *On Politics: A History of Political Thought from Herodotus to the Present*

Roberto Calasso, *La Folie Baudelaire*

Carolyn Abbate and Roger Parker, *A History of Opera: The Last Four Hundred Years*

Yang Jisheng, *Tombstone: The Untold Story of Mao's Great Famine*

Caleb Scharf, *Gravity's Engines: The Other Side of Black Holes*

Jancis Robinson, Julia Harding and José Vouillamoz, *Wine Grapes: A Complete Guide to 1,368 Vine Varieties, including their Origins and Flavours*

David Bownes, Oliver Green and Sam Mullins, *Underground: How the Tube Shaped London*

Niall Ferguson, *The Great Degeneration: How Institutions Decay and Economies Die*

Chrystia Freeland, *Plutocrats: The Rise of the New Global Super-Rich*

David Thomson, *The Big Screen: The Story of the Movies and What They Did to Us*

Halik Kochanski, *The Eagle Unbowed: Poland and the Poles in the Second World War*

Kofi Annan with Nader Mousavizadeh, *Interventions: A Life in War and Peace*

Mark Mazower, *Governing the World: The History of an Idea*

Anne Applebaum, *Iron Curtain: The Crushing of Eastern Europe 1944-56*

Steven Johnson, *Future Perfect: The Case for Progress in a Networked Age*

Christopher Clark, *The Sleepwalkers: How Europe Went to War in 1914*

Neil MacGregor, *Shakespeare's Restless World*

Nate Silver, *The Signal and the Noise: The Art and Science of Prediction*

Chinua Achebe, *There Was a Country: A Personal History of Biafra*

John Darwin, *Unfinished Empire: The Global Expansion of Britain*

Jerry Brotton, *A History of the World in Twelve Maps*

Patrick Hennessey, *KANDAK: Fighting with Afghans*

Katherine Angel, *Unmastered: A Book on Desire, Most Difficult to Tell*

David Priestland, *Merchant, Soldier, Sage: A New History of Power*

Stephen Alford, *The Watchers: A Secret History of the Reign of Elizabeth I*

Tom Feiling, *Short Walks from Bogotá: Journeys in the New Colombia*

Pankaj Mishra, *From the Ruins of Empire: The Revolt Against the West and the Remaking of Asia*

Geza Vermes, *Christian Beginnings: From Nazareth to Nicaea, AD 30-325*

Steve Coll, *Private Empire: ExxonMobil and American Power*

Joseph Stiglitz, *The Price of Inequality*

Dambisa Moyo, *Winner Take All: China's Race for Resources and What it Means for Us*

Robert Skidelsky and Edward Skidelsky, *How Much is Enough? The Love of Money, and the Case for the Good Life*

Frances Ashcroft, *The Spark of Life: Electricity in the Human Body*

Sebastian Seung, *Connectome: How the Brain's Wiring Makes Us Who We Are*

Callum Roberts, *Ocean of Life*

Orlando Figes, *Just Send Me Word: A True Story of Love and Survival in the Gulag*

Leonard Mlodinow, *Subliminal: The Revolution of the New Unconscious and What it Teaches Us about Ourselves*

John Romer, *A History of Ancient Egypt: From the First Farmers to the Great Pyramid*

Ruchir Sharma, *Breakout Nations: In Pursuit of the Next Economic Miracle*

Michael J. Sandel, *What Money Can't Buy: The Moral Limits of Markets*

Dominic Sandbrook, *Seasons in the Sun: The Battle for Britain, 1974-1979*

Tariq Ramadan, *The Arab Awakening: Islam and the New Middle East*

Jonathan Haidt, *The Righteous Mind: Why Good People are Divided by Politics and Religion*

Ahmed Rashid, *Pakistan on the Brink: The Future of Pakistan, Afghanistan and the West*

Tim Weiner, *Enemies: A History of the FBI*

Mark Pagel, *Wired for Culture: The Natural History of Human Cooperation*

George Dyson, *Turing's Cathedral: The Origins of the Digital Universe*

Cullen Murphy, *God's Jury: The Inquisition and the Making of the Modern World*

Richard Sennett, *Together: The Rituals, Pleasures and Politics of Co-operation*

Faramerz Dabhoiwala, *The Origins of Sex: A History of the First Sexual Revolution*

Roy F. Baumeister and John Tierney, *Willpower: Rediscovering Our Greatest Strength*

Jesse J. Prinz, *Beyond Human Nature: How Culture and Experience Shape Our Lives*

Robert Holland, *Blue-Water Empire: The British in the Mediterranean since 1800*

Jodi Kantor, *The Obamas: A Mission, A Marriage*

Philip Coggan, *Paper Promises: Money, Debt and the New World Order*

Charles Nicholl, *Traces Remain: Essays and Explorations*

Daniel Kahneman, *Thinking, Fast and Slow*

Hunter S. Thompson, *Fear and Loathing at Rolling Stone: The Essential Writing of Hunter S. Thompson*

Duncan Campbell-Smith, *Masters of the Post: The Authorized History of the Royal Mail*

Colin McEvedy, *Cities of the Classical World: An Atlas and Gazetteer of 120 Centres of Ancient Civilization*

Heike B. Görtemaker, *Eva Braun: Life with Hitler*

Brian Cox and Jeff Forshaw, *The Quantum Universe: Everything that Can Happen Does Happen*

Nathan D. Wolfe, *The Viral Storm: The Dawn of a New Pandemic Age*

Norman Davies, *Vanished Kingdoms: The History of Half-Forgotten Europe*

Michael Lewis, *Boomerang: The Meltdown Tour*

Steven Pinker, *The Better Angels of Our Nature: The Decline of Violence in History and Its Causes*

Robert Trivers, *Deceit and Self-Deception: Fooling Yourself the Better to Fool Others*

Thomas Penn, *Winter King: The Dawn of Tudor England*

Daniel Yergin, *The Quest: Energy, Security and the Remaking of the Modern World*

Michael Moore, *Here Comes Trouble: Stories from My Life*

Ali Soufan, *The Black Banners: Inside the Hunt for Al Qaeda*

Jason Burke, *The 9/11 Wars*

Timothy D. Wilson, *Redirect: The Surprising New Science of Psychological Change*

Ian Kershaw, *The End: Hitler's Germany, 1944-45*

T M Devine, *To the Ends of the Earth: Scotland's Global Diaspora, 1750-2010*